$$E = mc^2 = \sqrt{m_0^2 c^4 + p^2 c^2}$$

the equations
icons of knowledge

THE EQUATIONS

icons of knowledge

Sander Bais

 University Press 2005

ge, Massachusetts

The author wishes to thank his colleagues at the Institute for Theoretical Physics in Amsterdam and in particular Dr Leendert Suttorp for constructive comments and suggestions. Part of the work was done while visiting the Santa Fe Institute; its hospitality and stimulating atmosphere are gratefully acknowledged. The author is also indebted to the collaborators on the project at AUP, in particular Vanessa Nijweide, for their advice, enthusiasm and patience.

The biographical information came from various sources, such as:

Concise Dictionary of Scientific Biography, Charles Scribners and Sons, New York, 1981.

The Mac Tutor History of Mathematics Archive at:
www-groups.dcs.st-and.ac.uk/history

www.nobel.org

www.scienceworld.wolfram.com

Colophon

Design and lay-out Gijs Mathijs Ontwerpers, Amsterdam

This publication is printed on WOODSTOCK Camoscio by Fedrigoni

All rights reserved

First published in the Netherlands by Amsterdam University Press, Amsterdam

Library of Congress Cataloging-in-Publication Data

Bais, Sander.
The equations : icons of knowledge/Sander Bais.
p. cm.
ISBN 0-674-01967-9
1. Equations, Theory of. 2. Logic, Symbolic and mathematical. I. Title

QA211.B16 2005
512.9'4--dc22

Printed in Italy

Copyright © 2005 by Amsterdam University Press

Edition for the Netherlands, Amsterdam University Press: ISBN 90 5356 744 5

stal, but also dynamical processes like the scat-   8
ticles or the time evolution of any system we
interested in, be it the weather, the flow of a
vth of competing populations of bacteria or the
the universe. Equations often have predictive
olutions have pointed the way to new experi-
eries.

constants

nknowns (variables), equations usually contain
neters. These are constants, and by changing
one can drastically influence the properties of
that are obtained. Water may freeze or evapo-
ng on the temperature as a parameter. An orbit
rom bounded to unbounded, depending on the
e force of attraction.

nental equations we will focus on in this book,
rsal constants appear, often denoted as the
constants of nature', such as the gravitational
ewton, setting the strength of the gravitational
ocity of light, or Planck's constant. The values
nportant fundamental constants are given in a
aside of the front cover. These are key numbers
cale of things in our universe. If we would take
tions but change the values of these constants
n nature would certainly be very different. The
d well have been uninhabitable, and probably
have been here to discuss the matter. At this
e is no understanding of *why* the values are

Contents

Mathematics as a language of nature

This book is about the fundamental equations of the physical sciences as inspiring fruits of the human quest for understanding the universe. These equations are compact statements about the way nature works, expressed in the language of mathematics. As such, the equations we are going to discuss cannot be derived based on logical reasoning alone: they have resulted from a critical dialogue between the observation of nature and the intuition and creative thinking of some great minds.

Yet, this book does not aim at teaching either mathematics or physics in exact terms. We present the equations as plainly as possible, without much comment on the empirical data justifying them – despite the fact that their existence was revealed to us primarily through the careful and critical observation of patterns and motions in nature. This book is an attempt to convey the excitement and beauty of what these equations tell us.

Physicists use mathematics as a language of nature, a language which has had to be extended regularly when new layers of physical reality were uncovered. Indeed, the natural scientist tends to use mathematics as a language, whereas the mathematician studies it for its own sake.

Expressing relations by relating expressions

The word equation comes from the Latin *aequare*, which in turn is derived from the word *aequus*, meaning equal or level. Equations may only assign a value to some variable, but in our general context they express relations between physical variables that characterize the physical system we want to consider and determine their allowed changes in space and time. The relations are mathematically expressed using rela-

Introduction

A fashionable dogma

There is a fashionable dogma about popularizing science, which imposes a veto on the use of equations in any popular exposé of science. Some people hate equations, while others love them. The veto is like asking somebody to explain art without showing pictures. In this book, we override the veto and basically turn it around; for a change, the equations themselves will be the focus of attention. And if we talk about equations as *icons of knowledge*, we should certainly show them in all their beauty.

tional symbols such as the e[...] and less than '<' symbols. Co[...] tions or inequalities, dependi[...] appears.

There are many equations in[...] degree of importance. The [...] focus on mark radical turnin[...] They are the fundamental n[...] ideas embedded in a space of[...]

Simple equations and compl[...]

To study a system or struct[...] number of variables like posit[...] ature, and so on. We then tr[...] these structural variables, an[...] quite complicated. The variab[...] that we want to solve the equ[...] On all scales nature has org[...] that exhibit an astonishing s[...] complexity.

Simple laws can very well de[...] miracle is not the complexity[...] of the equations describing th[...] Our quest is twofold: one is[...] govern the structure and dy[...] while the other is to understa[...] and to obtain their solutio[...] requires knowledge of the sys[...] very much the task of the nat[...] ing the solutions the mathem[...] ble partner.

The solutions describe not on[...]

Law and order

The universe is highly ordered on all scales. On large scales, we can think of planetary systems that revolve around a star like our sun, of stars gathering in galaxies and of galaxies forming clusters or superclusters. In the micro world, we can think of atoms, nuclei and the most elementary building blocks of matter such as quarks, electrons and photons. In between, we find an immense diversity of complex structures in condensed forms of matter, like fluids, crystals and sand piles, as well as the perplexing structures of the molecules of life, such as DNA and proteins. Indeed, the systems we may want to study are of many sorts. They may also include population dynamics, the stock market, learning processes, epidemics or even the universe as a whole.

ecule or a cu[...] tering of pa[...] happen to [...] river, the gr[...] expansion [...] power: new[...] mental disc[...]

Variables a[...]

Besides the[...] control pa[...] their value[...] the solutio[...] rate, deper[...] may chang[...] strength o[...] In the func[...] certain un[...] 'fundamer[...] constant o[...] force; the [...] of the mos[...] table on th[...] that set th[...] the same e[...] of nature, [...] universe c[...] we would [...] moment [...] what they[...]

How this book could be read

My advice is to read through the introductory sections in which some key mathematical concepts and the symbols representing them are introduced rather casually. These sections introduce some of the vocabulary used extensively in the later sections, so you may even want to return to them from time to time.

The sections are not numbered, indeed indicating that the book does not have to be read from cover to cover. It is a kind of gallery in which you can wander from one room to another.

Finally, our equations know of each other. They embrace each other, walk hand in hand, or have hidden or manifest conflicts. Sometimes two of them get into a fight, and a third one emerges. Frequently, we run into what look like contradictions, which subsequently have to be resolved by introducing new unifying concepts – the crisis as creative moment. The equations tell us the story of paradigm shifts in science.

A bird's view

The content of this book is a kind of landscape, with the equations as mountains. Some of the mountains are hard to climb, but once on top the view is magnificent. We are going about it the easy way. We'll fly over the landscape getting glimpses of the highest peaks, not worrying about how at ground level it may be very hard to move from one place to another. We don't see the narrow passes and icy glaciers, the crevasses and steep walls. We skip the 99% sweat, as much as possible, to enjoy the 1% inspiration (to paraphrase Einstein's dictum on science). We do not walk the rocky paths; we see them as a network of thin threads laid over the landscape.

In choosing the bird's-eye view, we will not fully understand it all. For some it may be the first journey into this unknown territory. Beautiful poetry in a foreign language. It doesn't really matter, as long as one returns home with much more than one left with. This is why we give a map of contents before embarking on the trip.

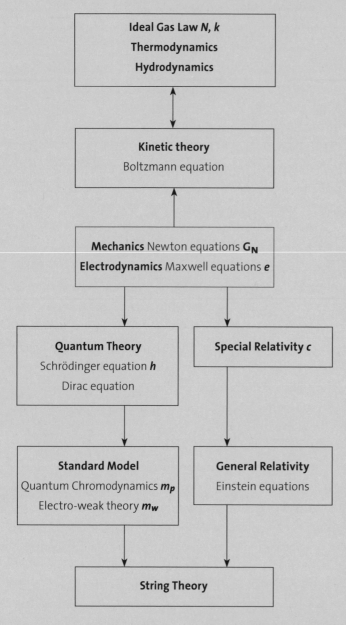

Map of contents

In the map of contents we have indicated some of the more obvious connections between the subjects. The arrows mostly indicate the chronology of discovery. In this book we will start in the middle and first move upward to the macroscopic world of materials consisting of very many atoms. After that we move down along the relativity and the quantum strands.

In the end we can think of the map having all arrows pointing upward, because that would reflect the structural hierarchy that is immanent in our world, and moreover, that's how nature unfolded itself during its 13.7 billion years of evolution since the Big Bang. This evolution at large, totally overarching Darwin's earthbound vision, demonstrates the unreasonable success of reductionism in science. The fact that Nature is becoming able to understand its own deepest origins may well mean that in our era we are going over an essential evolutionary threshold.

There is a certain hierarchy in the structure of equations. Their degree of mathematical complexity depends on the type of things one wants to describe. In high school most of us have been confronted with equations of some sort or another.

The tautological toolkit

The simplest equations are algebraic equations which only involve algebraic manipulations such as addition, multiplication and so on.

A simple algebraic equation that plays an important role in physics is the *equation of state* of an ideal gas. It is usually written as $PV=RT$ and expresses the phenomenological relation between the pressure P, the volume of a body of gas V, its temperature T and the fundamental gas constant R, which is just a known number.

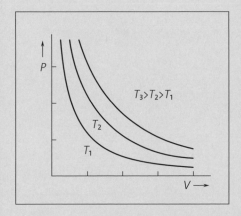

Such an equation can be used in many ways. It asserts that the three variables P, V, and T that characterize the state of a fixed amount of gas are not independent, because they have to satisfy the given relation. One obvious use of the equation is that if we have the actual values of two of the three variables, we can calculate the third variable. But the relation gives a lot more information of a qualitative kind (see figure). It follows, for example, that if one increases the pressure P while keeping the volume constant, then the temperature has to go up as well. Similarly, one may conclude that decreasing the volume while keeping the temperature constant, we have to go along one of the curves increasing the pressure.

$$PV = RT$$

You may be familiar with the way an equation like this one can be rewritten in different but equivalent forms, using the recipe: 'What you do to the left hand side, you've got to do to the right hand side.' So by subtracting RT from both sides we may write $PV-RT=0$, or by dividing both sides by T we get $PV/T=R$. Depending on the question one wants to answer, one

The gas constant R is given by $R=Nk$, where N is Avogadro's number, which specifies the number of molecules we are talking about. Boltzmann's constant k is a fundamental constant, relating temperature to energy.

or the other form may be more convenient. Even though the
equation may look different, the message remains the same.
Most of the equations we will discuss are not of this simple
algebraic type, but use more sophisticated notions like *derivatives*. Therefore, before embarking on the story of the fundamental equations, we have to introduce some of the frequently
recurring mathematical symbols and explain their meaning.

The language of quantities: variables, functions and fields

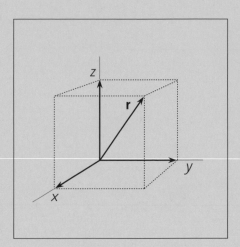

Let us now turn to some of the mathematical habits and symbols used by scientists. It is customary to denote quantities with
letters – temperature T, time t, position $\mathbf{r}$, etc. – and these quantities can take on different values and are therefore denoted as
variables. Certain constants also appear in the equations: on the
one hand there are *universal constants* like the velocity of light
c, or the charge of the electron e, and on the other hand there
will be other *parameters* which take on fixed numerical values in
a given context. An example is the viscosity η, which of course
depends on the type of medium you are considering: syrup is
more viscous than water. Instead of writing the numerical values
of the parameters in the equations, we also denote them by letters. So we distinguish several sets of letters in our equations:
variables, universal constants and system parameters.

Let us now add other levels to the notion of variables. Variables may depend on each other, in which case we refer to
them as a *function* or a *field*. For example, the temperature
in a room may depend on the position in the room and on
the time. We express this as $T(\mathbf{r},t)$, where the notation implies
that we think of temperature T as a function of space and
time. A function like T embodies a lot of knowledge, because
you only know it completely if you know its value for all points
in space and all moments in time.

To make life even more complicated, variables and functions may have different *components*: they are rather a collection of things. The different components are indicated by an index (usually in sub- or superscript), labeling which member of the collection one is referring to. For example, the position **r** in three dimensions has three components, with $r_1 = x$, $r_2 = y$ and $r_3 = z$. The combined position **r** is called a *vector*. The convention is that variables or functions that have components are printed in bold. The boldface letter is like the family name, while the components indicated by an index are like the members of the family indicated by their initials.

The word component refers to some reference frame, which in most cases of interest to us will be a rectangular or Cartesian coordinate system. Roughly speaking, if we want to indicate where we are – say, in an office in Manhattan – we have to specify three numbers: one for the avenue (x), one for the street (y) and one to indicate on which floor we are (z).

Vectors can be defined in any dimension, and in general they have a length and a direction. For example, **r** can be represented as an arrow pointing from a chosen reference point called the *origin* with coordinates (0,0,0) to the point with coordinates (x,y,z)(see figure page 12). This way of thinking about vectors anticipates their use. If we rotate a vector, its direction changes but its length stays the same. Collections of variables carry fancy names like vectors, tensors, multiplets, spinors, etc., which refer to their specific properties, such as how they behave under certain operations like a rotation.

Finally, we add one more level of complexity by combining the previous notions. We may for example consider a function or field with components, like the 'velocity field' of water in a river **v**(x,y,z,t): it is a field of vectors that describes what the components of the velocity vector **v** = (v_1,v_2,v_3) are as func-

Inversions

If we have defined a mathematical operation, it always pays off to wonder what the inverse operation is. Operations tend to come in pairs that are able to neutralize each other. In defining the appropriate inversions, important mathematical discoveries have been made.

Playing with marbles one learns to count, to add the positive integers. If one loses enough of them in a game, one is painfully led to the idea of subtraction, and perhaps discovers the all-important number zero. Having nothing left, one may shift to a more abstract level and try to subtract even larger numbers, and so one may invent the negative integers. Similarly, multiplication leads to division, and thus to the discovery of the rational numbers. By taking roots of positive numbers, one is led to the continuous set of the real numbers, whereas roots of negative numbers lead to the introduction of complex numbers. Sometimes small questions have great answers.

tions of space and time, i.e. $v_1 = v_1(x,y,z,t)$, etc. Alternatively, one may think of such a *vector field* as a map that assigns a vector **v** to any space-time point (x,y,z,t).

A useful way to think about a vector field is to imagine a water flow: at any point there is a little vector telling you what the velocity of the water at that point is. The water particles stream along certain streamlines which cannot intersect. To draw the velocity field at a certain point, one draws an arrow tangent to the flow line through that point, while its length is proportional to the density of the flow lines in the neighborhood of that point. So we are used to drawing a vector field at a certain time as a pattern of field lines through space (see figure).

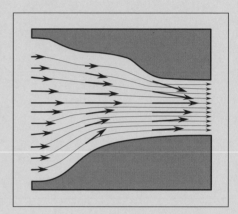

The language of changes: derivatives

The functions or fields for a fixed moment in time describe a situation (that is, the state of the system) at that given moment. Most things, however, vary in space and time. Equations often determine how quantities can evolve from a given initial state. They also specify how the variations of the variables in space and time depend on each other. Newton's laws, for example, describe how the position and velocity of a particle change, given some external force. Therefore, these *dynamical equations* are often called 'equations of motion'.

The laws of hydrodynamics for instance describe how fluid density, fluid velocity, and the temperature of a fluid change in space and time, depending on the forces present in the fluid as well as on certain typical constants that characterize the fluid, such as its viscosity and thermal conductivity. What we need is a language of changes that belongs to a particular branch of mathematics called *differential calculus*. This calculus was developed independently by Isaac Newton and Gottfried Wilhelm von Leibniz at the end of seventeenth century.

Their tools made it possible to describe dynamical systems in precise mathematical terms, and therefore marked the beginning of the quantitative understanding and prediction of natural phenomena.

As the word suggests, *differentials* are nothing more than infinitesimally small changes. The symbols used in *differential equations* are rather straightforward. For a particle in motion, its position vector **r** will depend on time, so $\mathbf{r} = \mathbf{r}(t)$; the infinitesimal change of the position in time is called its *time derivative* and denoted as $d\mathbf{r}/dt$. The time derivative basically tells us the infinitesimal change in the position **r** (denoted as $d\mathbf{r}$), *per* infinitesimal change in time dt. It tells us which way the position tends to change at that point – the 'trend' in position, as it were. The 'differential quotient' $d\mathbf{r}/dt$ is therefore nothing more than the velocity **v** of the particle at a certain instant, and is given in units of meters *per* second, as suggested by the notation.

We may write that the acceleration **a** of an object is the time derivative of the velocity, expressed as $\mathbf{a} = d\mathbf{v}/dt$. Note that both velocity and acceleration are vectors (they have a direction as well as a magnitude), and that in general they will depend on time. It is quite straightforward to generalize to higher derivatives: for example, the second derivative is nothing but the derivative of a derivative. For the case on hand, one may say that the acceleration is the second derivative of the position with respect to time, which is now simply denoted as $\mathbf{a} = d^2\mathbf{r}/dt^2$, meaning that you take the time derivative twice. If the derivative at a point t in time tells us something about the trend of the function in the vicinity of that point, the second derivative tells us something about the 'trend of trends' around that point.

Is there a simple way to visualize these changes? Yes, if one thinks about it graphically, one can, for example, visualize the motion of a particle as a curve showing the points it is traveling through, by plotting the time horizontally and the distance vertically (see figure). The time derivative of the position at a certain time t_0, in other words the velocity at time t_0, is nothing but the slope of the curve at time t_0 (see figure). The derivative of a function gives the 'steepness' or 'slope' of the curve, which can be positive or negative. If the curve is horizontal, as in a local maximum or minimum, the velocity (and hence the derivative of the function) is zero. A second derivative should be thought of as the curve's curvature at a certain point. For example, negative acceleration (deceleration) would correspond to the line curving down.

This symbolic way of thinking about changes and variations is simple, logical, and elegant indeed, and maybe most importantly, extremely useful. As mentioned before, most of the equations we are going to discuss are differential equations or, in other words, equations in which derivatives appear. They are different from the so-called algebraic equations most of us are familiar with (such as the quadratic equation $ax^2+bx+c = 0$, where the problem is to solve x given a, b, and c).

A simple example is the celebrated formula of Newton, $\mathbf{F} = m\mathbf{a}$, which can also be written as $\mathbf{a} = \mathbf{F}/m$. And since $\mathbf{a} = d\mathbf{v}/dt$, we get $d\mathbf{v}/dt = \mathbf{F}/m$. This is a differential equation which has to be solved for the velocity $\mathbf{v} = \mathbf{v}(t)$ as a function of $\mathbf{F}/m$, which, in principle, can be dependent on x and t as well.

Functions may depend on more than a single variable, and then the function may change depending on those variables. The figure shows a function f of two variables x and y. It is clear that at an arbitrary point P on the surface, the slope in the x-direction may differ from the slope in the y-direction. An example of such a function is the temperature, which varies smoothly from place to place and from time to time.

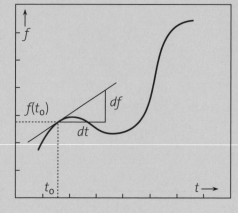

Trading in derivatives

The notion of derivatives also exists in the world of finance. It refers to financial products like options, futures and swaps that derive their security from underlying primary financial products like stocks, commodities, exchange rates, etc. They are a kind of insurance against unpredictably large fluctuations that are a threat to the holder, but of course they are at the same time used to speculate. These derivatives are a different trade from what we discuss here, except that both are dealing with 'trends'.

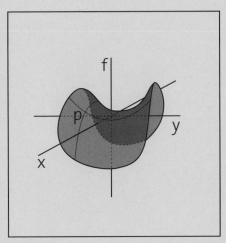

This fact requires the introduction not only of time derivatives but also of spatial derivatives d/dx, d/dy and d/dz – and in cases where there are more possible dependencies we use curly or partial derivatives $\partial/\partial x$, $\partial/\partial y$, $\partial/\partial z$. These three spatial derivatives (giving the slopes of a function in three different directions) may then be combined in a vector and represented by a single symbol, ∇, denoted as the *nabla* or *del* operator. The nabla operator gives the slopes in the different directions. To say it more precisely: applying it to a function $f(x,y,z)$ yields a vector field $\mathbf{g}(x,y,z)$ with components $g_1 = \partial f/\partial x$, $g_2 = \partial f/\partial y$, $g_3 = \partial f/\partial z$.

Adding up the differences: integration
To solve a differential equation, we need to know how to do the opposite of taking the derivative. If we first take the derivative of a function and then apply this operation, we should get back the function we started with. If we are given the value of a function g at a given instant t_0 and also the values of its derivatives for all times t, we can reconstruct the function by adding up all the increments - the differences dg. In the language of continuous differences, mathematicians have given a precise definition of this notion and called it *integration*. This very heuristic description of the concept of integration suffices for this book, as we will never have to explicitly integrate the equations.

The nabla symbol – an inverted delta – refers to an ancient triangular Assyrian harp. The name 'nabla' was coined by Tait, who got into a lighthearted dispute with Maxwell about the matter. Maxwell preferred the term 'slope' for the operator.

This concludes our lightning review of some of the vocabulary and symbols used in the equations. Most of it is also summarized in a table at the inside back cover.

The growth of lilies in a pond; the increase of a population of bacteria, rats or people; the spreading of a contagious disease or of an ideal chain letter: these are all examples of exponential growth. The scary thing about exponential growth is that numbers get really big very fast. Exponential growth processes are doomed to terminate, because of a shortage of something that is needed to keep the process going, such as space, food or money.

The opposite of growth is decay. From a mathematical point of view, decay is very similar to growth: all it takes is changing a sign in the equation. Because the population depends only on time, these processes are described by what are called *ordinary differential equations*, which contain a derivative with respect to only one variable – in this case time. We look first at the simple equation for unlimited growth and decay, and then at the *logistic equation*, in which the growth saturates at a certain equilibrium value.

The simplest version of a growth equation states that the change in time of a quantity $n(t)$ is proportional to its magnitude at that time. Say, $n(t)$ denotes the number of mice in a cheese factory at a given time t. These mice have ample food and will start multiplying like mad. This means that the total reproduction at any given time will be proportional to the number of mice at that instant. In other words, the change per unit of time of n – i.e. the time derivative dn/dt – will be proportional to n. Translated into mathematics this yields the differential equation given in the side bar, with r the reproduction rate or (Malthusian) growth rate (r is larger than zero).

The solution for $n(t)$ to this differential equation is an exponential growth of the number with time, as given on

Rise and fall
The logistic equation

$$\frac{dn}{dt} = rn$$

$$n(t) = n_0 e^{rt}$$

Historical note

The Belgian mathematician Pierre Verhulst, inspired by the work of Malthus, proposed the logistic equation for limited growth in 1838. He became politically engaged after the Belgian revolution of 1830 and the invasion of his country by the Dutch army in 1831. Though his political commitment did not bear fruit, he pursued his interests in social issues as a mathematician and a teacher.

Verhulst's interest in probability theory was triggered by a new lottery game, but under influence of Malthus's theory of population growth, he soon applied it to political economy and later to demographical studies. Due to his fragile health, he died in 1849 at the age of 45.

$$\frac{dn}{dt} = rn\left(1 - \frac{n}{k}\right)$$

page 18 with n_0 the number of mice at time zero. The fact that the time appears in the exponent, together with the positive constant r, implies that the growth will be extremely rapid. This is illustrated by the steeply rising dashed curve in the figure on page 21.

If we choose a negative number for the rate r, we have an equation for an exponential decay. It can be used to describe, for example, the number of radioactive nuclei in a chunk of radioactive material as a function of time.

Returning to the mice population, we should not be content with the model described: it tells only part of the story. After all, the solution of this equation suggests an unlimited growth, which is not very realistic. At a certain moment the mouse population will have grown to such an extent that the limited cheese supply will lead to a food shortage, which will effectively reduce the growth rate. The growth may come to a halt when n reaches a certain critical value, which we will call k. To take this effect into account, one has to change the equation, and one arrives at the so-called *logistic equation* proposed by Pierre Verhulst in 1838.

The figure shows the population n as a function of t for various initial values n_0. After a while (= for large values of t), the population approaches the limiting value k. This happens because as n gets closer to k, the expression on the right hand side of the equation keeps decreasing and therefore so does dn/dt.

Note that if we start with a situation where n_0 is larger than k, the sign on the right hand side becomes negative and therefore the derivative becomes negative, leading to negative growth (or decay) until the equilibrium value $n=k$ is reached.

The fall of an empire

There is a story about a starving Chinese beggar who made the humble request to the emperor to be given some rice according to the 'chessboard rule'. The first day the emperor would put two grains of rice on the first square, the second day four grains on the second square, the third day eight grains on the third square, etc. The poor emperor agreed, unaware that he was to ruin his empire. The number of rice grains doubles every day, and after 64 days he would have to give

$$2^{64} = 18,446,744,073,709,551,616$$

grains of rice, about a thousand billion tons of rice!

$$n(t)= \frac{k}{1+(k/n_0 -1)e^{-rt}}$$

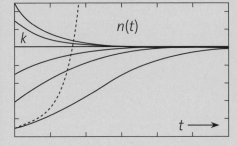

21   The exact formula for the solution is given in the side bar. More complicated cases of population dynamics as used in ecology, with for example p different competing species, can be described by a system of p coupled equations of the type just described. Such systems are Darwinistic in that they may incorporate the feature of *natural selection*.

The potential threat of unlimited population growth was first put on the agenda by Thomas Robert Malthus (1766–1834), who was renowned for his pessimistic predictions regarding the future of humanity. He made his major contribution to economic thought in the essay *The Principle of Population*. Originally, Malthus wrote the piece in response to utopian utilitarians who suggested that population growth constituted an unmitigated blessing.

Essentially, Malthus predicted that the demand for food inevitably becomes much greater than the supply of it. This prediction was rooted in the idea that the population increases exponentially with time, while food production only grows as the time raised to some power. Not surprisingly, Malthus's ideas had a considerable influence on Darwin and his idea of natural selection. Things have worked out differently from what he predicted because of improved production and fertilisation techniques, yet the problem of global population growth has certainly not been solved and has regained the urgency first put forward by Malthus.

Newton's three dynamical equations describe the motion of a body with mass m under the influence of a force F. They form the heart of Newton's principal work, the *Principia Mathematica*, where for the first time motion is defined and described in precise mathematical terms using derivatives. These equations were the starting point for the mathematical modeling of dynamical systems in the most general sense. They form the cradle of quantitative science. The fourth equation gives the explicit expression for the gravitational force. Together the equations provided the explanation of the gravitational phenomena – both celestial and terrestrial – known in Newton's time. These laws furnished the solid theoretical foundation for the findings of Copernicus, Brahe, Kepler, and Galileo concerning the elliptic orbital motion of the planets around the sun, but they also accounted for why an apple falls from a tree. They explain how different forces balance each other to guarantee the stability of bridges and buildings, and on the other hand why constructions sometimes collapse.

These equations involve time derivatives, because the velocity $v = d\mathbf{r}/dt$ is the change in time of position and the acceleration $\mathbf{a} = d\mathbf{v}/dt$ is the change in time of velocity. Position, velocity and acceleration are all vectors: they have a magnitude and point in a certain direction.

The first equation defines what the *momentum* $\mathbf{p}$ is; this is sometimes called 'the amount of motion'.

The second, most celebrated equation determines the motion in terms of the force exerted on a body. If we apply a force $\mathbf{F}$, this causes an acceleration $\mathbf{a}$ in the same direction as the force. The equation implies in particular that if no force is exerted, there will be no change in velocity, and the momentum will

Mechanics and gravity
Newton's dynamical equations and universal law of gravity

Isaac Newton

Isaac Newton was born on Christmas Day in 1642 in a small manor house in the village of Woolsthorpe, England. At the age of twelve, he was sent to King's School at Grantham, where he didn't stand out as a genius. He left home in 1660 to prepare for Cambridge and subsequently went to Trinity College a year later. Within eight years, he had accepted the Lucasian Chair of Mathematics. It took another fifteen years before he began writing his principal work, *Principia Mathematica*, a work of true genius that changed the notion of science forever.

Though received with enthusiasm, Newton's work was not fully accepted and taught in universities until after his death. In 1699, he was appointed Master of the Mint in London. In 1727, Newton, the scientist, administrator, mystic, and theologian – he wrote thousands of pages about theology – died at his home in Kensington. He was buried in Westminster Abbey.

$$\mathbf{p} = m\,\mathbf{v}$$

$$\mathbf{F} = m\,\mathbf{a}$$

$$\mathbf{F}_{1 \rightarrow 2} = -\,\mathbf{F}_{2 \rightarrow 1}$$

$$F_G = G_N\,\frac{m_1 m_2}{r^2}$$

remain constant. In the early days of mechanics, it was quite
a revolutionary idea that motion would persist in the absence
of a force – a rather disturbing, counter-intuitive thought,
moreover in conflict with the commonly held views of
Aristotle. If there are many forces working on a particle – for
example friction, air resistance, and an electric force – then **F**
is the net resulting force that the particle experiences.

The third equation states that if we exert a force on a body,
then that body will exert an equal but opposite force on us; in
short, 'action is reaction'. The sun exerts a force on the earth,
and the earth exerts the same force, but in the opposite
direction, on the sun. A book exerts a force on the table, but
the table exerts the same force on the book, otherwise it
would fall right through.

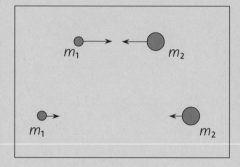

The last equation gives the formula for the gravitational
force. This law is strikingly simple: it states that any two
objects with masses m_1 and m_2 attract each other with a
force $\mathbf{F}_G$, which is proportional to the product of the masses
and inversely proportional to the square of the distance r
between them. Indeed, at twice the distance, the strength
of the attraction becomes one-fourth (see figure). The
proportionality constant G_N, not surprisingly called Newton's
constant, is universal in the sense that it is the same for all
bodies of mass, regardless of what material they are made
of. Together with the equations of motion, the gravitational
equation gave a quantitative explanation for the planetary
motions around the sun, but it also correctly described
gravitational effects on earth, notably the observation made
by Galileo that the acceleration g of freely falling objects on
earth does *not* depend on the mass of the objects. This law is
obtained from Newton's equations if one replaces m_1 by the
earth's mass M, and r by its radius R. If we then compare the

result with Galileo's formula $F=mg$, we obtain the acceleration $g = G_N M/R^2$. The force law in effect unified the description of terrestrial and celestial mechanics.

These four simple laws were a victory of mind over matter. What makes the application range of the laws of Newton so wide is the fact that the force **F** can be any force, it doesn't have to be gravity. Newton's equations allow us, at least in principle, to calculate the motion of many interesting systems: cannonballs, satellites and planets, but also cars, violin strings, pendulums, hula-hoops, windmills, swimmers, bicycles and bungee jumpers. For static considerations, too – to calculate the distribution of forces acting in bridges or buildings, for example – the equations are an indispensable starting point.

Conservation of energy

Energy conservation is a cherished principle respected in all of physics, but to properly understand it one must include all possible forms of energy such as heat, radiation, binding energy and eventually also mass. For simple systems like the motion of an object under a position-dependent force (like a swing, or any motion in a gravitational field, etc.) the expression for the total energy U is given by:

$$U = \frac{1}{2}mv^2 + V(x)$$

The first term which contains the velocity squared, is the kinetic part related to the motion, whereas the second term $V(x)$ corresponds to the potential energy. The force is defined as minus the derivative of $V(x)$, or $F(x) = -dV(x)/dx$. If the potential energy becomes steep, then the force gets large. It follows from Newton's equations that the total energy is conserved: $dU/dt = 0$. It was precisely through our understanding of these simple systems that the fundamental importance of notions like energy became apparent.

Nonlinear dynamics and chaos

Towards the end of the nineteenth century the French mathematical physicist Henri Poincaré pointed out that even relatively simple systems obeying innocuous-looking differential equations like Newton's laws can have orbits that are extremely complicated. And what is more important, changing the initial conditions very slightly can cause the motion to become very different. This extreme sensitivity to the initial conditions has important consequences: it leads to what is called chaotic behavior. If we repeat a real experiment or a computer simulation thereof, we can only reproduce the initial conditions to a certain finite accuracy. Clearly, if the system we study has an extreme sensitivity to initial conditions, the subsequent outcomes of such 'identical' experiments would be entirely different. This severely upsets our ability to predict exactly what will happen. This behavior of effective unpredictability in a strictly speaking well-defined, completely deterministic system is called deterministic chaos. This type of chaos is a very generic feature of nonlinear dynamical systems and has been a major subject of investigation since the last quarter of the previous century. Chaotic behavior is not linked to any specific equation; it refers to the type of behavior that solutions of nonlinear equations may exhibit if the values of the parameters in the equations lie in a certain range.

A particularly simple example is a 'gravitational billiard' studied by Lehtihet and Miller. The system consists of a ball in a gravitational field, moving in a vertical plane. It is bounced completely elastically by the walls of a wedge shaped container (see figure). The trajectories for the ball between the walls can easily be calculated using Newton's dynamical equations combined with the gravitational force described in the previous sections. The ball experiences a constant vertical acceleration g downward due to the gravitational force and therefore describes an orbit, which between bounces is just a segment of some parabola (just like the motion of a ball in the earth's gravitational field). When it bounces, the component of the velocity perpendicular to the wall reverses direction. In spite of the apparent simplicity of this system, the ball's motion may be changed from a simple periodic motion to chaotic motion by just changing the initial condition or the opening angle.

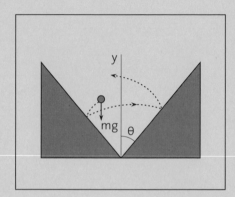

Newton's heritage

If one writes down the equations for a system of objects, including all of the forces between them, we obtain a closed set of coupled differential equations, which describes the dynamics of the whole system. These features are generic for large classes of very different dynamical systems, and indeed this is why Newton's *Principia* paved the way for the quantitative scientific description of virtually any dynamical system. In mechanics, the variables are the positions and velocities of particles, but these could be population densities of various species in an ecological system or production factors such as prices, stocks, and wages in an economic model. As mentioned, Newtonian dynamical systems, though deterministic, include a wide variety of chaotic systems. Simple systems – like two bodies that attract one another through a central gravitational force – can be solved exactly in terms of well-known mathematical expressions and functions. However, for most problems no exact solutions are known, and one has to rely on a more qualitative global analysis or on approximate computer-generated solutions to gain insight into what the solutions look like and what their properties are.

The braid of science

Modern science is not just a dialogue between theory and experiment, often also the computer plays an important and quite independent role in studying the equations and analyzing the behavior of their solutions. Experiment, theory and computer simulation form a perfect *ménage à trois*.

Newton's dynamical equations hold for any force that may be applied to an object. Newton himself gave the expression for the gravitational force. Of the other forces in nature, the electromagnetic force is of paramount importance, because it is the force that keeps atoms and molecules together. To a large extent this force is responsible for the properties of ordinary matter in all its diverse manifestations, from solids to nerve cells. Basic to all these manifestations is the understanding of the force exerted on a charged particle when it moves through a given external electric and magnetic field. Hendrik Antoon Lorentz established the electromagnetic force law in its general form.

The electromagnetic force
The Lorentz force law

The Lorentz force law contains two contributions. The first term in the equation tells us that an arbitrary (positive) charge q moving in an electric field feels a force in the same direction as the electric field **E**. This means that if the electric field is uniform in space and constant in time, the force exerted on the charged particle will be constant, and therefore the particle will move with a constant acceleration in the direction of the electric field. A big charge feels a stronger force than a small charge, and will feel a stronger acceleration if it has the same mass. Indeed, electric fields can be used to accelerate particles. This force also leads to the attraction between two oppositely charged particles: the one charge feels the force of the electric field caused by the other, which may result in them forming a bound pair.

Now the second term, describing the force resulting from a magnetic field **B**, is of a very different nature. Note that the force is again proportional to the charge q, but also to the velocity **v** of the particle. Maybe surprisingly, this means that

Hendrik Antoon Lorentz

Lorentz was born in 1853 in Arnhem, the Netherlands, where his father ran a flourishing gardening business. He entered Leiden University at the age of 16. In 1875, at the early age of 22, Lorentz obtained his doctor's degree, and only two years later he was appointed to the Chair of Theoretical Physics at Leiden, newly created for him.

There he managed to combine atomistic views with the Maxwell equations, but he also stood at the threshold of the theory of relativity. His enormous international reputation was based on his profound knowledge and judgment, which he combined with a patient and modest character. In 1902, he shared one of the first Nobel Prizes with Pieter Zeeman for his explanation of the line splitting in atomic spectra as observed by Zeeman. Lorentz died in 1928 in Haarlem.

$$\mathbf{F}_{em} = q\mathbf{E} + \frac{q\mathbf{v}}{c} \times \mathbf{B}$$

a particle at rest does not feel the magnetic field. The 'cross' product of the two vectors **v** and **B** means that the force is perpendicular to both the velocity of the charge and the magnetic field, as indicated in the figure. The magnitude of the cross product equals the product of the magnitudes of **v** and **B** times the sine of the angle between them. So if the two vectors are parallel, their cross product is zero (because sin 0 = 0).

So what does this magnetic force do? If the particle moves through a constant magnetic field, then it will feel a constant force perpendicular to its velocity all the time. It will therefore undergo an acceleration of constant magnitude perpendicular to its velocity. If the velocity is perpendicular to the field to begin with, then the upshot is that the charge will move in a circular orbit in a plane perpendicular to the direction of the field. If the velocity is not perpendicular to the field, the component of the velocity in the direction of the field will not change, while the one perpendicular will cause a circular motion. Combining the two velocity components, one finds that a charge will move in a spiral motion around the magnetic field lines.

This is what happens with the charged particles that come to the earth from the sun. They are 'caught' by the magnetic field lines of the earth's magnetic field and spiral up to the magnetic poles, where the field is stronger. This causes the charges to radiate and light up the sky: the phenomenon known as *northern lights*.

Often one is not dealing with a single charge that moves, but with a continuous charge distribution that is spread over space, then in Lorentz's equation the charge q has to be replaced by the charge density ρ and the quantity $q\mathbf{v}$ by the current density **j**.

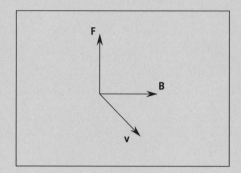

Multiplication of vectors

Adopting smart notational conventions ensures that the equations become maximally compact, making their structure very transparent. For two vectors **u** and **v** we may define two types of product: one is called the scalar (or dot) product because it yields a single number (scalar) and is written with a dot: $s = \mathbf{u} \cdot \mathbf{v} = uv \cos \varphi$, where u and v are the length of **u** and **v** and φ is the angle between them. So this means for example that $\mathbf{u} \cdot \mathbf{u} = u^2$. The other is called the vector (or cross) product because it is written with a cross and yields a vector: $\mathbf{w} = \mathbf{u} \times \mathbf{v}$. The vector **w** is pointing in a direction perpendicular to the plane spanned by **u** and **v**, and the length is given by $w = uv \sin \varphi$. This implies that $\mathbf{u} \times \mathbf{u} = 0$.

31 The question that remains at this point is: where do the electric and magnetic fields come from, and which law determines what they look like? These fields are caused by the positions and motions of charges, so what is needed in addition to Newton's equations and the Lorentz force law is a set of equations determining the electric and magnetic fields given a distribution of charges in space and time. Clearly, this has to be a very fundamental set of equations – once these are specified, we can combine them with the Newtonian equations to get a closed system of equations for the motions of charges and the changes in the fields. These two sets of equations with two sets of unknowns ideally would have to be solved simultaneously. But let us first turn to a preliminary equation, involving vectors and partial derivatives.

Water flows through rivers, gas through pipes, electric current through wires, traffic through cities, people through buildings and cash flows through our hands. Indeed, charge can flow and form an electric current, but charge cannot just disappear: it is conserved. There is always a distribution of charge over space and time, called its density. The *continuity equation* is the precise mathematical relation between a flow or current and a change in density, and is called a *conservation law*. This type of equation is called a *partial differential equation* because it involves time and spatial derivatives of functions of time and space.

A local conservation law
The continuity equation

The continuity equation is a flow equation that expresses a conservation law that is *local* because it applies to any given volume, on all scales. If in an office there are a certain number of people, that number can only change if people enter or leave (through the doors). This is true for any one room, but also for a floor, a whole building or a city.

The continuity equation basically states that the amount ρ (of charge, or gas, or money) in a given volume element can only change if there is a corresponding current or flow $\mathbf{j}$ going through the surface bounding the volume element: charge may flow, but it cannot disappear.

For a *stationary flow*[*] we have $\partial\rho/\partial t = 0$. This condition says that the 'fluid' is incompressible, because the local density ρ of the fluid does not change. For a stationary flow the equation says that for any volume, the same amount of fluid has to flow in and out. That's why a river streams faster through a narrow and shallow passage. This condition typically applies to a real liquid, but not to a gas or a stream of people. Many people may enter a room while nobody leaves, in which case the average ρ in the room would increase.

A slow flow

Pitch is a derivative of tar and appears to be solid at room temperature. It is even so brittle that it can be shattered with a blow of a hammer. Yet it is a fluid, but a very viscous fluid. There is an amusing experiment started in 1927 by Thomas Parnell of the University of Queensland. He put some pitch in a sealed glass funnel and removed the seal in 1930. The rest of the experiment consisted of waiting and watching. Every nine years a droplet came through the funnel: a very slow flow indeed. The experiment is still running.

[*] The word 'stationary' does not mean static: the flow is constant but there is still a flow, implying that something moves. It is like an engine running stationary.

$$\frac{\partial \rho}{\partial t} + \nabla \cdot \mathbf{j} = 0$$

The Maxwell equations stand at the very basis of the whole edifice of classical physics. Where Newton's laws tell us how particles move once the forces are specified, the Maxwell equations allow us to specify the electromagnetic forces. They determine the electric and magnetic fields, E and B respectively, caused by a given distribution of charges, or generally by a charge density ρ and a density of electrical currents j. Knowing the fields the resulting Lorentz force on charges can be calculated.

Maxwell's equations describe in full generality the electromagnetic phenomena. From these equations follows for instance the existence of electromagnetic radiation, which – depending on its wavelength – we know as radio waves, visible light or X-rays. It also follows that this radiation is emitted by charges if they get accelerated. This theory is a magnificent example of what is called a *field theory*, exhibiting the crucial property that fields of force like the electromagnetic fields represent physical degrees of freedom, which carry energy and momentum and which can propagate in space and time (in the form of radiation).

The equations at first look contrived, but in fact the form they are presented in is both natural and efficient, wonderfully pairing convenience with transparency. The expressions and derivative operators which feature in these equations are quite intricate, but basically follow naturally from the fact that we are considering vector fields. The electric and magnetic fields that appear in the Maxwell equations are vector fields, meaning that they depend on space and time, and furthermore they have three components each. So at any point in space there exist an electric and a magnetic field vector, each pointing in some direction in space. Time derivatives of the fields appear in two of the equations,

Electrodynamics
The Maxwell equations

James Clerk Maxwell

Electromagnetism has a long history of experimental and theoretical development; some names worth mentioning are Coulomb, Ampère and Faraday. We can say that Maxwell was the person who completed the monumental construct of the theory of electromagnetism.

Maxwell was born into an old Scottish family near Edinburgh in 1831. After holding professorships in Aberdeen and London, he withdrew to the countryside in 1865 to compose his *magnum opus* 'A Treatise on Electricity and Magnetism', which appeared in 1873. In it, all the details of classical electrodynamics were brought together in the four equations we present here. Thereafter, Maxwell also made essential contributions to the kinetic theory of gases. In 1871, he became the first director of the Cavendish Laboratory in Cambridge. This is where he died at the early age of 48.

$$\nabla \cdot \mathbf{E} = \rho$$

$$\nabla \times \mathbf{B} - \frac{1}{c}\frac{\partial \mathbf{E}}{\partial t} = \frac{\mathbf{j}}{c}$$

$$\nabla \cdot \mathbf{B} = 0$$

$$\nabla \times \mathbf{E} + \frac{1}{c}\frac{\partial \mathbf{B}}{\partial t} = 0$$

determining the dynamics of the electric and magnetic fields; those are the 'equations of motion'.

The first Maxwell equation determines the electric field caused by the presence of charges ρ and is known as Gauss' law. The form of the equation is very similar to the continuity equation discussed before. The charge density ρ is a source (or sink) of electric field lines. For example, if ρ corresponds to a positive charge q sitting at some point, the solution for **E** will be a purely spherically symmetric field that points radially out – away from the charge. This radial field gives rise to the Coulomb force between two charges, which is an inverse square law very similar to Newton's gravitational force law, except that the masses get replaced by charges and the proportionality constant reflecting the strength of the interaction is different.

The second equation – known as Ampère's law – is an equation that links electric and magnetic phenomena. It basically states that electric currents **j** (or moving charges) may cause magnetic fields around them, similar to the way static charges cause electric fields. In this equation we encounter the cross derivative of the magnetic field, in which the vector of spatial derivatives, denoted as nabla, is applied to the three components of the magnetic field vector **B**, and the current **j** acts as a source for that field.

Ampère's law explains the deflection of a compass needle, if we bring it close to a wire carrying an electric current. If **j** corresponds to a current through a thin wire along a straight line, there will be a magnetic field **B** circling around this line (see figure on page 37). But the second Maxwell equation also shows that changing electric fields can cause magnetic fields, because $\partial\mathbf{E}/\partial t$ will be non-zero in that case.

Comparing forces

We may compare the strength of the gravitational and electric forces by comparing their magnitude for a pair of fundamental particles like the electron. The ratio of the forces is independent of the distance between the two particles. So we can ask the question whether the electrons will fly apart due to the electric repulsion, or whether they will move towards each other due to their gravitational attraction? It turns out that the electric force is immensely stronger (in fact by some 39 orders of magnitude!). That the electromagnetic forces are much stronger than the gravitational force can also be shown by a simple experiment. Recall that a small magnet is strong enough to lift an iron nail off the ground, i.e. the small magnet exerts a force larger than the gravitational force, which is the combined effort of all matter in the entire earth beneath that nail.

The third equation looks somewhat like the first one, except that the electric field has been replaced by the magnetic field and the right-hand side is zero. The latter is a reflection of the basic fact that there are no such things as magnetic charges, by which we mean isolated magnetic North or South poles. If we take a bar magnet and break it into two pieces, we get two bar magnets, and if we go on breaking these into ever-smaller pieces, this continues to happen over and over again. The only things we find in nature are magnetic dipoles – bound pairs comprised of a North and a South pole. However, an electric current going around a coil or a current loop can also generate the field of a magnetic dipole (this actually follows from the second Maxwell equation). And indeed, it turns out that all observed magnetic phenomena can be understood as caused by electric charges moving around. Therefore, the third Maxwell equation stipulates the nonexistence of isolated magnetic poles. We see that the Maxwell equations possess a kind of symmetry between electric and magnetic properties, except for the charges and currents that occur in nature, which seem only to exist in the electric variety.

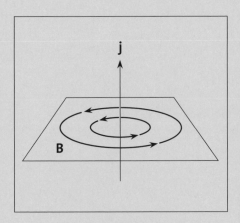

The final equation is usually referred to as Faraday's law; it expresses the fact that changing the magnetic field may induce an electric field, which in turn may cause charges to move. The charges start moving in such a way that they tend to counteract any change in the magnetic field by creating compensating magnetic fields.

Linearity

Looking at the Maxwell equations we see that the electromagnetic fields only appear linearly (there are no squares or higher powers of **E** and **B**). From this linearity follows the important property that one can simply add solutions. For example, if one has first solved the equations for some current and some charges separately, then the solution for the combined configuration containing both charges and current is just the sum of the two separate solutions.

The structure of this system of differential equations is obviously rather involved. It is a theory of fields, of vector fields even. One considers the charge and current densities as given. Then one must solve the equations for the **E** and **B** fields, which include three components each. Thus, one comes up with a total of six unknown field components. The first and third equations have a single component, while the second and fourth equations each include three components because they are vector equations having three components each. Thus, we can conclude that we have eight equations for six field components, which, generally speaking, would make the system over-determined. If a system of equations is over-determined, generally it is inconsistent. The correct interpretation of the situation here is as follows: the second and fourth equations are used for the time evolution of the fields, while the other two are merely used to relate the different field components to each other at any given time. In other words, they form a *constraint* on the field configuration at any one instant.

If that is in fact the case, there could in principle be a problem of inconsistency. If we take a field configuration that satisfies the constraint equations at, say, time zero, then we have to verify that by letting the configuration evolve according to the dynamical equations, the constraints should not be violated at any time. Maxwell showed this to always be the case for his system of equations. In fact, it made him add a term to the original Ampère's law, which in its original form would have rendered the system inconsistent.

One more question comes to mind. We see that both charges and currents feature in these equations. In the section on the continuity equation, we discussed the conservation of charges. So where did that equation go? The answer is simple; the Maxwell equations imply the continuity equation, so that the conservation of electric charge does not have to be added as an extra equation.

Assuming that no other forces are present, the motion of

Monopoles: to be or not to be?

So far, magnetic monopoles have never been observed. This of course does not rule out that they may still exist in principle. A strong argument in favor of their existence was given in 1931 by the physicist Paul Dirac. He concluded from combining Maxwell's electromagnetism with quantum theory that the mere existence of a single monopole would explain the quantization of electric charge, which is observed in nature: all free particles in nature have charges which are integer multiples of the electron charge. Maybe monopoles only occur in tightly bound pairs, or maybe North and South poles were annihilated in an early stage of the universe. In the theoretical arena of unified theories for elementary particles and fundamental forces, monopoles regularly make their comeback. So behind the scenes monopoles are still active.

Electromagnetic energy

The electromagnetic field carries certain energy. The energy of some field configuration can be obtained by adding up the different energy contributions of the little volume elements that make up the total volume. The jargon is to say that we integrate the energy density over the volume. What then is the energy function, or energy density? It is just given by:

$$U = \frac{1}{2}\mathbf{E}^2 + \frac{1}{2}\mathbf{B}^2$$

This expression has some similarity to the expression for the energy of a mechanical system, with the first term giving the kinetic energy and the second term the potential energy. The analogy with mechanical systems can be pushed even further, because the field has also a momentum density which can elegantly be expressed as a vector field:

$$\mathbf{S} = c\mathbf{E} \times \mathbf{B}$$

These formulas show how basic the notions of energy and momentum are.

charged particles in an external electromagnetic field is completely fixed by Lorentz's force law in combination with Newton's equations. The Maxwell equations in turn describe the electromagnetic fields caused by given external charges and currents. Now if we combine Newton's equations with the Maxwell equations, we obtain a closed system for the motion of the particles and the fields they give rise to. To make the story really consistent, one must in fact solve the Newtonian and Maxwell equations simultaneously, taking into account the back reaction that fields and charges/currents have on each other. This is accurately described by the complete set of equations.

Therefore, this set offers a unified description of electric and magnetic phenomena, from the forces between charges – as previously formulated by Coulomb – to systems like electromotors, dynamos, transformers, antennas, plasmas (fluids consisting of charged particles) and particle accelerators. And that is not all: as mentioned before, it is also the theory of electromagnetic radiation, such as light. A pleasant surprise, to which we turn in the following section.

Waves are a very generic phenomenon: water waves, sound waves, radio waves and fashion cycles are all periodic motions of some sort. Waves are oscillations that propagate through a medium such as water or air, or even through empty space, as is the case with the electromagnetic waves described by the equations given here. These wave equations follow directly from the Maxwell equations in the absence of charges and currents. We exhibit them separately because of their tremendous importance.

Electromagnetic waves
The wave equations

Propagating waves are characterized by a wavelength λ, a velocity v and a frequency f. Between these three quantities there exists a simple relation, namely $\lambda f = v$. The magnitude of the oscillation is called the *amplitude*. The electromagnetic waves involve propagating electric and magnetic fields, with the property that the magnetic and electric oscillations are perpendicular to the direction of propagation and perpendicular to each other – so-called transversal waves (see figure).

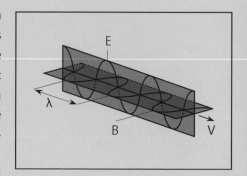

Maxwell showed that the propagation speed of these waves exactly equals the known speed of light c. Visible light was apparently just an electromagnetic wave phenomenon and turned out to be part of a continuous spectrum of electromagnetic waves, which vary in wavelength and frequency. Moving from very long to very short wavelengths, we identify this radiation subsequently with radio waves, microwaves, heat radiation, infrared light, visible light, ultraviolet light, and X-rays.

Maxwell's equations thus managed to unify the description of electric, magnetic, optical and radiation phenomena. The combined equations of Maxwell and Newton therefore form the beating heart of all of classical physics.

The ether
Electromagnetic waves were believed to propagate through some medium called 'ether' that filled all of space. It was only after Einstein had formulated his Special Relativity that the hypothetical ether was abolished, and it was understood that electromagnetic waves just propagate through empty space.

$$\frac{1}{c^2}\frac{\partial^2 \mathbf{E}}{\partial t^2} - \nabla^2\,\mathbf{E} = 0$$

$$\frac{1}{c^2}\frac{\partial^2 \mathbf{B}}{\partial t^2} - \nabla^2\,\mathbf{B} = 0$$

The study of solitary waves or *solitons* started in the nineteenth century when a young Scottish engineer, John Scott-Russell, observed a single wave which, guided through a local canal, seemed to travel as far as he could see without losing its shape. He jumped on a horse, so the story goes, and followed the wave for miles along the canal. A soliton is just a single bump that does not broaden, lose its shape or weaken as it travels along in a medium (see figure).

The stability of solitons is truly remarkable, and this makes them very relevant for applications, varying from error-free transmission of light pulses through a glass fiber, or electric pulses through nerve cells, to understanding tsunamis.

For a long time there was a fierce debate involving many scientific celebrities about whether solitary waves could really exist; because of the natural tendency of waves to broaden and dissipate, the phenomenon required a miraculous conspiracy. The debate was settled when Korteweg and his student De Vries wrote down their equation for water waves in a shallow rectangular canal in 1895, and rigorously showed that the equation did indeed have solitary wave solutions. The equation is nonlinear because the function *u* appears quadratic in the last term, indeed this equation is very different from the linear equations for electromagnetic waves, yet it is the prototype of large families of nonlinear soliton equations which share similar properties.

Solitons have other remarkable properties, for example, their speed of propagation is proportional to their amplitude or height – so a big wave will overtake a small one. Furthermore, if solitons meet and interact, they will eventually re-emerge unaffected and continue their trip with a slight time delay. So these waves are not able to destroy each other, and that is a property which makes them extremely useful.

Solitary waves
The Korteweg–De Vries equation

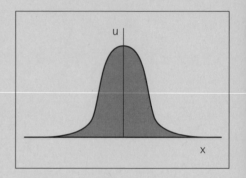

Diederik Johannes Korteweg and Gustav de Vries

Korteweg was born in 1848 in 's Hertogenbosch. After his studies in Delft, he moved to Amsterdam to work on a doctoral thesis under Van der Waals. Three years later, in 1881, Korteweg was appointed a professor of mathematics, mechanics and astronomy at the University of Amsterdam. His research in fluid dynamics led him to the formulation of the famous equation, which appeared for the first time in the doctoral thesis of De Vries in 1895. Korteweg died at the age of 93 in 1941.

$$\frac{\partial u}{\partial t} + \frac{\partial^3 u}{\partial x^3} + 6u\,\frac{\partial u}{\partial x} = 0$$

The three laws of thermodynamics describe systems that exchange energy with each other or with some environment. This exchange may be the system doing work or heat being absorbed. In principle, the number of particles could also be changed. These processes may be quasi-static, so that the condition of equilibrium is maintained during the process and the process is reversible, but it may also be that the system goes from one equilibrium state to another via a non-equilibrium process (for example, the free expansion of a gas). The 'executive summary' consists of the following statements:

1 The first law states that heat is a form of energy and that energy is conserved.
2 The second law tells us that a system cannot convert all absorbed heat into work. Machines that are one hundred percent efficient do not exist. In the second law a new important state variable, the entropy S, is introduced.
3 There is a lowest temperature, at which a system is maximally ordered, and where the entropy trends to zero. This is a consequence of the quantum-mechanical nature of any physical system, which becomes relevant at very low temperatures.

It is striking that states of macroscopic systems consisting of very many particles which are in equilibrium can be described effectively in terms of a very small number of variables and parameters. This is quite a general phenomenon, holding for large classes of liquids, gases and solids, and mixtures thereof.

In thermodynamics we are interested in describing processes of macroscopic (sub)systems in which some kind of energy exchange takes place. That may be because we bring them into

Thermodynamics
The three laws of thermodynamics

Historical note

Thermodynamics arose before the atomic nature of matter was understood. The German physician Julius Robert Mayer (1814–1878) was the first who stated explicitly in 1842 that heat is a form of energy. Recognition of his great discovery and the applications found by Mayer came only very much later, and this made him bitter and desperate. Work on heat engines goes back to the work of Nicolas Léonard Sadi Carnot (1796–1832), who formulated the law of entropy which became known as the second law of thermodynamics. Thermodynamics as a consistent theory was formulated by Clausius and Lord Kelvin, and was further developed by Josiah Willard Gibbs around 1875. The third law, which indicates that entropy vanishes for zero absolute temperature, is due to Hermann Walther Nernst (1864–1941).

$$dU = đQ - đW$$

$$dS = \frac{đQ}{T} \qquad \frac{dS}{dt} \geq 0$$

$$T \to 0 \quad \Rightarrow \quad S \to 0$$

thermal contact, or we exert forces on them, etc. Now we will be concerned with changes in the variables that characterize the state of the system. The number of independent variables is equal to the number of ways that energy can be supplied or extracted from the system. Each of these variables is paired with a dependent variable. Half of these state variables are extensive (i.e. proportional to the amount of material or the size of the system), like for example the volume V, and the other half are intensive, like P and T. So to completely describe the thermodynamical properties, one needs an even number of variables, while in the equation of state (like $PV=RT$, as we discussed in the introductory sections) we typically only have an odd number. This suggests that we expect to encounter yet another thermodynamic variable.

As we are now interested in changes of the state of the system, it is not surprising that again we will be using differentials and derivatives to describe the infinitesimal changes that the system may undergo.

The first law of thermodynamics basically states that heat is a form of energy. More precisely, it says that the change in energy dU equals the heat absorbed by the system dQ minus the amount of work done by the system dW. Alternatively, one could say that the first law states the *conservation of energy*. The equation suggests two special cases: one where there is no work done ($dW = 0$), and the other where no heat is exchanged ($dQ = 0$), in which case we speak of an *adiabatic* process.

Now here we have to be careful, as not all processes that are energetically allowed do occur. Sure enough, it is possible to convert work entirely into heat, but whether it also works the other way around is far from obvious.

We all know that if a wheel is heated up, it will not start turning around all by itself – though this would be allowed by energy conservation. It's not impossible but highly, indeed highly improbable. On the other hand, a steam engine does convert heat into work. The important question 'To what extent can heat be converted to work?' is answered by the second law of thermodynamics, undoubtedly the most famous of the three. This law featured in the famous essay 'The Two Cultures' by C.P. Snow, published in *The New Statesman* in 1956, in which Snow publicly criticized the scientific illiteracy of well educated person by noting the fact that virtually everybody knows a play by Shakespeare but virtually nobody knows what the – equally important – second law of thermodynamics is about!

There are two different but equivalent formulations of the second law. The Kelvin formulation states that there is no thermodynamic process whose sole effect is to extract heat from a reservoir and to convert it entirely into work. This can be paraphrased by saying that an ideal engine cannot exist as a matter of principle. This has to be contrasted with a *real* engine, which is a machine that goes through a cycle of thermodynamic states and does convert heat into work, but at the same time has to deliver part of the extracted heat to the environment (which is a reservoir at a lower temperature). It cannot convert all heat that is extracted into work, which means that such a real engine is never one hundred percent efficient.

The Clausius formulation of the second law says that there is no thermodynamic process whose sole effect is to extract heat from a colder reservoir and to deliver it to a hotter reservoir. In other words, an ideal refrigerator does not exist. Indeed, in a real refrigerator work has to be done to move heat

Mission impossible

Inventors of all nations have spent an enormous amount of work to construct a so-called *perpetuum mobile*, which is a machine that once put in motion keeps going forever without an external supply of energy. A device that itself generates the energy needed to keep it running. The most ingenious contraptions have been proposed. Some are extremely efficient. However, the impossibility of a perpetual mobile can be rigorously proved from the first two laws of thermodynamics. The French Academy of Sciences decided in 1775 to no longer consider proposals or publications in which the existence of a perpetuum mobile was claimed. In spite of this, new proposals are still popping up from time to time.

from the colder to the hotter reservoir; as everybody knows, it takes energy to refrigerate.

The second law involves the notion of *entropy*, denoted by S. It is a state variable like temperature, but more subtle because one cannot measure it directly. The change in entropy is related to the temperature T and heat dQ. One may say that an infinitesimal amount of heat absorbed by the system leads to a change in the entropy of the system given by the formula $dQ = TdS$.

In terms of entropy, the second law of thermodynamics states that for any process taking place in a closed system, the entropy cannot decrease. For example, for the ideal refrigerator a (positive) amount of heat q would be extracted from the system at a low temperature T_1 and subsequently that same amount q would be delivered to the environment at a higher temperature T_2. The total entropy change $\triangle S$ would be $\triangle S = -q/T_1 + q/T_2$, which equals $\triangle S = q(1/T_2 - 1/T_1)$, and that would be smaller than 0, violating the second law. The conclusion is therefore that an ideal refrigerator cannot exist.

Heat tends to cause random motions in the system – the molecules start colliding with more energy – so one might say that the notion of entropy is a measure for the *disorder* in the system. This definition can be made rigorous if one studies the statistical connection with the accessible states of the system at the microscopic level. In the field of statistical mechanics, one can establish that in a closed system which is not in equilibrium, the entropy increases until it reaches a maximum at equilibrium. This is basically a consequence of the trend to 'go from a less probable to a more probable state'. If a drop of ink is released in a container filled with

Entropy and information

On the microscopic level entropy can be defined as the logarithm of the number of accessible states, and because raising the temperature makes more states accessible, the entropy and the disorder increase. This definition of entropy is the same as the definition of information in modern information theory starting with the work of Claude Elwood Shannon. So the information content of a book is the logarithm of all possible ways the book could have been filled with words (or with letters).

How can information be the same as disorder? Imagine two classrooms full of children. In one classroom there is a lot of discipline and the all children sing 'Happy birthday'. In the other, the teacher went away and left the children by themselves, the children are all talking at the same time. Clearly, the disordered class produces about 25 times as much information as the 'single birthday voice', so, it does make sense to identify disorder and information.

49 water, the ink molecules will scatter around and spread out in the water until they have reached the equilibrium situation where the molecules are distributed uniformly throughout the container. Clearly, the opposite process will not happen, certainly not spontaneously. Think of the following analogy: if one randomly starts repositioning objects in a teenager's room, one definitely will not create order but a mess. And as many of us know too well, this is the equilibrium state in which most teenagers' rooms end up. To create order, one has to do the opposite and make a specific effort to put things exactly in the place where they belong. This indeed takes quite some energy.

The considerations just mentioned have led to the suggestive statement that since entropy – and therefore the amount of disorder – has to increase, the ultimate fate of any closed system is utter chaos, the so-called 'heat death'. This would presumably also apply to the universe as a whole – a rather dim perspective indeed. However, we should keep in mind that the second law does not prohibit that within a closed system a certain subsystem may decrease its entropy by giving off heat, which then has to be absorbed by other parts of the total system. That is the way the increasing order and complexity are achieved, for example, in biological systems.

The third law of thermodynamics explains what happens if a system approaches the lowest possible temperature of zero Kelvin, about -273° Celsius. At this temperature thermal motion will stop, and the system will get maximally ordered, the number of accessible states is reduced to its lowest energy state, and if this state is unique, the entropy should become zero. The existence of such a state follows from the quantum properties of nature.

The Boltzmann equation forms the basis of kinetic theory, which provides the bridge between microscopic and macroscopic physics. This equation belongs to the realm of *statistical mechanics*, and it describes the motion of large numbers of particles in a statistical sense. The Boltzmann equation allows for a systematic derivation of macroscopic transport processes like diffusion, heat flow and conductivity from the underlying microscopic laws of nature. Though it applies only to dilute systems, it underlies theories that are of tremendous practical and technological importance, such as fluid, aero- and plasma dynamics. This equation forms the solid basis for the description of systems that are not in equilibrium, in particular for analyzing processes that lead to equilibrium.

Kinetic theory
The Boltzmann equation

Historical note

Clausius, Maxwell and Boltzmann pioneered the microscopic approach to macroscopic phenomena. Ludwig Boltzmann (1844–1906) formulated the equation carrying his name in 1872. Systematic methods to solve this equation were developed by Chapman and Enskog around 1916.

Boltzmann was an energetic and impulsive character. He studied in Vienna, Austria, and by the age of 25 had already become professor of Physics in Graz. There he spent his most productive years. During this period he suffered from the lack of understanding and recognition of his contemporaries. The story goes that Boltzmann smuggled a bottle of wine into the Berkeley Faculty Club while visiting the school in 1904, when Berkeley was a 'dry' town. At the age of 62 he committed suicide.

To describe the behavior of macroscopic systems like gases or liquids, one does not need to know the detailed properties of all the individual particles making up the gas or liquid. Fortunately, because that would involve solving a system of 10^{23} or more coupled equations. It suffices to know the average properties of particles, and that is where statistical considerations come in. Now the fact that we have to deal with very many particles becomes a blessing in disguise: as every casino owner or insurance company can tell you, statistical arguments become extremely accurate and thus powerful if one is dealing with large numbers. So, whereas the motion of the individual particles in a fluid or gas may be quite random, the statistical distribution function describing the probabilities of where they are and what velocity they have is not, and it is this distribution function which has to satisfy some fundamental equation. In a gas, for example, evidently not all molecules sit together in one place: they spread more or less

$$\left(\frac{\partial}{\partial t} + \mathbf{v} \cdot \nabla_{\mathbf{r}} + \frac{\mathbf{F}}{m} \cdot \nabla_{\mathbf{v}}\right) f =$$

$$\int d\Omega \int d\mathbf{v}_1 \sigma(\Omega) |\mathbf{v} - \mathbf{v}_1| (f' f'_1 - f f_1)$$

evenly through the available space, and one also expects that because of the collisions of the molecules, the energy will be more or less equally distributed among the molecules once the system is in equilibrium. The importance of the Boltzmann equation is that it also describes processes that deviate from the equilibrium; for example, it allows one to prove that a process leads to equilibrium.

More precisely, the Boltzmann equation describes the time evolution of the distribution function $f(\mathbf{r},\mathbf{v},t)$ which specifies the probability of a particle being found at a position $\mathbf{r}$ with a velocity $\mathbf{v}$ at some time t in a gas, fluid or plasma. The equation involves derivatives with respect to all the variables – time, space and velocity – on which the function f depends. The left-hand side of this equation describes the change in time of the distribution as a consequence of the free flow of the fluid – in other words, because of the velocity of the particles and the external force that may be applied to the system. The complicated expression on the right-hand side approximates the effect of the collisions between the particles on the change of the distribution function. This interaction depends on the relative velocity and the scattering cross-section σ, which in turn depends on the relative scattering angle Ω. Its precise and complicated form need not concern us here.

To get some feeling for what this rather abstract distribution function means: one may use it to define certain averages which yield more familiar physical quantities. For example, if we integrate the distribution function over all velocities, we obtain the 'number density' $n(\mathbf{r},t)$ of particles which depends only on space and time. The number density gives the probability to find a particle at position $\mathbf{r}$ at time t, irrespective of the velocity it has.

Emergent properties

Water is made up of a very large number of molecules which interact with each other. This huge collective displays beautiful properties, like waves, vortices etc. These are properties of the medium that the constituents of the medium have not. Water molecules have not the faintest idea of what water waves are! These are therefore called *emergent properties*. As there are many layers of structure below what we see with the naked eye, there is in fact a whole hierarchy of emergent phenomena. We emphasize however that there are no extra interactions or influences introduced at any level; emergent phenomena are just the result of a subtle interplay between the underlying physical degrees of freedom and the fundamental forces between them. There is mounting evidence that even life is an emergent phenomenon.

$$n(\mathbf{r},t) = \int f(\mathbf{r},\mathbf{v},t)\, d^3\mathbf{v}$$

$$f(\mathbf{v}) = n\left(\frac{m}{2\pi kT}\right)^{\frac{3}{2}} e^{\frac{m|\mathbf{v}-\mathbf{u}|^2}{2kT}}$$

There is a simple case for which the solution for the distribution function has a particular importance. If we assume that the system is in equilibrium, then the distribution is per definition time-independent and homogeneous (which means independent of position), and if we furthermore assume that there is no external force working, then the left-hand side of the Boltzmann equation would vanish for any function that depends on the velocity only. The right-hand side is in general only zero if f depends in a suitable way on any quantity which is conserved in two-particle collision processes, such as mass, momentum and energy. From these assumptions Maxwell constructed an explicit solution for the equilibrium distribution of the velocities v around their average value u. This is called a Gaussian distribution: a symmetric function sharply peaked around v=u that falls off very rapidly if v deviates from u. From these assumptions Boltzmann constructed the expression (which was earlier obtained by Maxwell using different arguments) for the equilibrium distribution of the velocities **v** around their average value **u**.

Hydrodynamics is used to describe the flow of continuous media like gases, many types of fluids and plasmas. It is used in computer modeling to optimize the design of ships, airplanes and sports cars with respect to stability and resistance. The equations are also used for weather prediction, including computer simulations of hurricanes, tornadoes and tsunamis.

The three equations of hydrodynamics are basically conservation laws. They can describe a wide variety of transport phenomena because the equations contain quite a few parameters. These equations were originally obtained through the study of fluids. The system can also be obtained from the Boltzmann equation by analyzing averages of quantities that are conserved in two-particle collisions: mass, momentum and energy.

A very hard problem still facing the mathematical and physical community is to understand in detail the phenomenon of *turbulence* from first principles, and to construct it as a solution to the Navier–Stokes equations.

The Navier–Stokes equations are based on the conservation of three basic quantities in the underlying particle interactions: mass, momentum and energy. Related to these are three fields: the mass density $\rho(r,t)$, the velocity field $u(r,t)$ and the energy density (per unit mass) $\varepsilon(r,t)$. The equations require knowledge of the equations of state of the medium one is studying, one relating the pressure to density and temperature, $P = P(\rho,T)$, and one relating the energy density to density and temperature, $\varepsilon = \varepsilon(\rho,T)$.

For dilute gaseous systems these fields are defined as averages over the velocities using the Boltzmann distribution function, as given in the side bar on page 57. The defining expressions

Hydrodynamics
The Navier–Stokes equations

Navier and Stokes

Although Newton and Bernoulli considered problems involving fluids, it was the Swiss mathematician Leonhard Euler (1707–1783) who laid down the foundations of hydrodynamics through a systematic investigation of its basic equations around 1755. It is only appropriate that the 'inviscid equation' of motion is named after Euler, who is often regarded as the originator of modern hydrodynamics.

Claude-Louis Navier (France, 1785–1836) was the son of a lawyer who died when he was 8 years old. He was left to the care of an uncle, a civil engineer, who tried to stimulate his interest in engineering. Navier made a poor start at the École Polytechnique, but after two years was among the best students. He entered the École des Ponts et Chaussées in 1804 and graduated two years later. He became a famous builder of bridges, though the first bridge he built apparently collapsed.

$$\frac{\partial \rho}{\partial t} + \nabla \cdot (\rho \boldsymbol{u}) = 0$$

$$\left(\frac{\partial}{\partial t} + \boldsymbol{u} \cdot \nabla\right)\boldsymbol{u} = \frac{\boldsymbol{F}}{m} - \frac{1}{\rho}\nabla\left(P - \frac{\eta}{3}\nabla \cdot \boldsymbol{u}\right) + \frac{\eta}{\rho}\nabla^2 \boldsymbol{u}$$

$$\left(\frac{\partial}{\partial t} + \boldsymbol{u} \cdot \nabla\right)\epsilon = -\frac{P}{\rho}\nabla \cdot \boldsymbol{u} + \frac{K}{\rho}\nabla^2 T$$

for the fields depend only on **r** and *t*. One may then use the Boltzmann equation to derive the set of coupled equations given above, usually called the Navier-Stokes equations.

It is a very general set of equations, and in principle they also describe aerodynamics, though the parameters will be very different. In these equations a number of approximations have been made by introducing several phenomenological parameters which characterize the fluid, a pressure tensor P, the viscosity η, and the thermal conductivity K. We have left out a viscosity-dependent contribution to the third equation for simplicity.

Note that the first equation is really just the continuity equation for the fluid. The second equation reduces to an equation first given by Euler if the viscosity is set to zero. What one learns from the Boltzmann-equation approach to such a phenomenological system of macroscopic equations is that the phenomenological parameters themselves can be understood as certain averages over microscopic degrees of freedom, and corrections can in principle be calculated in a systematic approach.

We see that such equations take a rather complex form, yet they are nothing but the expression of a number of rather simple conservation laws applied to a many-particle system. The left-hand sides represent just the transport due to the free convection of the flow. The right-hand sides contain the effects of stress and viscosity, giving rise to complex diffusive behavior of the gas or fluid which may lead to a turbulent flow.

The equations can be used to describe fluid flows through pipes, in rivers and around ships or airplane wings. For special cases considerable simplifications to the equations

Nowadays Navier is remembered for the equations that carry his name. The story is basically that he arrived at the correct equations by the wrong arguments in the year 1822. In 1831 he became a professor at the École Polytechnique. He was politically quite active, believed strongly in the blessings of an industrialized society and opposed the bloodshed of the French revolution and the military expansionism of Napoleon.

George Gabriel Stokes (1819–1903) was the son of the Rector of Sligo College in Skreen, Ireland. He studied in Cambridge and was the eldest of a group of renowned theoreticians to which also W. Thomson (Lord Kelvin) and Lord Rayleigh belonged. Stokes occupied the Lucasian chair and was both president and secretary of the Royal Society (the latter for not less than 30 years!). His most important work is that on hydrodynamics.

are possible. For example, the Bernoulli equation relating pressure and velocity for a stationary fluid flow can be easily derived, as can the equation for the propagation of sound waves, because these are just longitudinal density waves. Upon setting the velocity field u equal to zero, the energy equation basically reduces to the diffusion equation for heat flow.

$$\rho(\mathbf{r},t) = m \int f(\mathbf{r},\mathbf{v},t) d^3\mathbf{v} = mn(\mathbf{r},t)$$

$$\rho(\mathbf{r},t)\mathbf{u}(\mathbf{r},t) = m \int \mathbf{v} f(\mathbf{r},\mathbf{v},t) d^3\mathbf{v}$$

$$\rho(r,t)\epsilon(r,t) = \frac{1}{2}m \int |\mathbf{v}-\mathbf{u}|^2 f(\mathbf{r},\mathbf{v},t) d^3\mathbf{v}$$

Einstein published his Theory of Special Relativity in 1905 and the General Theory of Relativity, covered in the next section, about ten years later. Special relativity describes how one's perception of space and time depends on the way one is moving. The notion of an absolute time, present in the Newtonian view of nature, turned out to be untenable. These fundamental differences become dramatic if one considers objects that move with relative velocities close to the speed of light.

The equations given here highlight three striking consequences. Firstly, the fact that velocities add up in a peculiar way such that one can never reach a velocity greater than the speed of light. This speed limit is universal, in the sense that it is the same for all observers. Secondly, the fact that moving clocks tick slower, quantifying the basic idea that time is relative. Finally, we highlight the most celebrated equation of twentieth-century physics, which expresses the fundamental equivalence of mass and energy.

Special relativity is based on two fundamental postulates concerning different observers who travel with a constant speed with respect to each other; so-called 'inertial observers'. The first is that the laws of physics should be the same for all inertial observers – for all of them the equations should look exactly the same. The second is that for all inertial observers the velocity of light (in vacuum) is the same.

The first postulate may not be so surprising: it is the relativity postulate and in fact not so new. Galileo already formulated it in a rather explicit way. The postulate boils down to saying that if I do experiments to discover the laws of physics on the ground, and somebody else does the same experiment in a train that moves with constant speed, then we will obtain the same laws of physics. It entails that if you are traveling with a constant

Special relativity
Relativistic kinematics

Albert Einstein

Albert Einstein was born in Ulm, Germany, on March 14, 1879. He went to the Luitpold Gymnasium in Munich, but he didn't like the regimentation of school. Later, after his family moved to Milan, he continued his education at Aarau in Switzerland. In 1896 Einstein entered the Swiss Federal Polytechnic School in Zürich, hoping to become a physics and mathematics teacher.

In 1901, the year he gained his diploma, Einstein acquired Swiss citizenship. As he was unable to find a teaching job, he accepted a position as technical assistant in the Swiss Patent Office. He devoted his spare time to problems in theoretical physics. In 1905 he obtained his doctor's degree. That same year he published three seminal papers in *Die Annalen der Physik* on three completely different problems: the photoelectric effect, Brownian

$$w = \frac{u+v}{1+\dfrac{uv}{c^2}}$$

$$t' = t\sqrt{1-\frac{v^2}{c^2}}$$

$$E = mc^2 = \sqrt{m_0^2 c^4 + p^2 c^2}$$

speed with respect to somebody else, there is no objective way to decide who is moving and who is not. This is something you may in fact have experienced while sitting in a slowly moving train. The notion of motion is relative.

The second postulate is certainly surprising, and even when understood remains counter-intuitive. It is a consequence of the Maxwell equations for the propagation of light waves. Think of being on a train traveling with a speed u, and throwing a ball with velocity v to a person on the platform. Our daily life experience, or what we like to call 'common sense', would tell us that the person on the platform would receive the ball with a velocity w which is the sum of the two velocities, so $w = u+v$. That is indeed what Newton would have told us. But if we now replace the ball with a particle of light – a photon – then the second Einstein postulate plainly says that the photon, which moves with respect to the train with the velocity of light, will move with exactly the same speed with respect to the person on the platform. How strange! The answer Einstein gave to the question of how two velocities (in the same direction) 'add' is provided by the first equation of the three. It is in fact a rather simple algebraic formula. Note that we can immediately read off some special cases. First, if both u and v are much smaller than the speed of light c, then uv/c^2 is very much smaller than one, and the Einstein formula reduces to Newton's formula $w = u+v$, as it should. If $u = c/2$ and $v = c/2$, however, Einstein's first formula tells us that $w = 4c/5$. And finally if $v = c$, then w is also c for all values of u, as required by the second postulate.

This formula tells us something very fundamental about velocities. Observe that even if we keep adding an arbitrary number of velocities smaller or equal to c, we can never get a velocity larger than c! Apparently the velocity of light is the maximum allowed velocity in nature, and fair enough (by the second postulate),

motion and Special Relativity. Seven years after starting at the patent office he was appointed Privatdozent in Bern. In 1909 he became an associate professor in Zürich, in 1911 Professor of Theoretical Physics in Prague, returning to Zürich in the following year to fill a similar post. In 1914 he was appointed Director of the Kaiser Wilhelm Physical Institute and Professor at the University of Berlin. He became a German citizen in 1914 and remained in Berlin until 1933, when he renounced his citizenship for obvious political reasons. Einstein received the Nobel Prize in Physics in 1921 'for his services to Theoretical Physics, and especially for his discovery of the law of the photoelectric effect', remarkably not for his work on relativity. Whereas in 1901 Einstein considered giving up his academic aspirations, in 1933 he received offers from many places like Jerusalem, Leiden, Oxford, Madrid and

that maximum is the same for all observers. As far as we know this is the speed limit that nature obeys without exception.

This absolute bound on the velocity also applies necessarily to all conceivable ways information can be transferred, which is crucial in saving the extremely fundamental concept of causality – the rules of cause and effect – on which not only science but also daily life is based. With propagation speeds higher than that of light, one could undo actions that took place in the past, which would lead to completely absurd and unacceptable consequences. Of course it would be a blessing if one could *a posteriori* prevent an accident which had already taken place, but it would also be hard to believe in such a miracle.

How can it be that the velocity of light is measured to be equal for all observers, one may ask. Well, the price is indeed quite high. Remember that velocity is distance divided by time, and that is where the clue lies. For the speed of light to be the same for all observers, the relations between their notions of space and time have to be redefined. In particular, the Newtonian idea of an absolute time turned out to be no longer tenable. Because the separation between what we call space and what we call time depends on how we move, it makes more sense to speak about space-time rather than space and time separately. Space-time is the collection of all space-time points (x,t), or of all 'events' as physicists like to call them.

The way space-time is experienced by different observers is completely determined by Einstein's theory. The adjective 'relative' means that not all observers perceive all events and the order in which they may occur identically, but the theory does allow you to determine what the differences are and how they come about.

One striking consequence is that moving clocks run slower,

Paris. He decided to go to the U.S. to take up an appointment at the Institute for Advanced Study in Princeton. He became a United States citizen in 1940 and retired from his post in 1945. In Princeton he worked mainly on a geometric theory unifying electromagnetism and gravity.

The price for his genius was his dwelling in intellectual solitude, though music also enriched his life considerably. He married Mileva Maric in 1903 and they had a daughter and two sons; their marriage was dissolved in 1919 after which he married his cousin, Elsa Löwenthal, who died in 1936. Einstein was an outspoken pacifist; his last letter was addressed to Bertrand Russell in which he agreed that his name should go on a manifesto urging all nations to give up nuclear weapons. Einstein was cremated at Trenton, New Jersey, at 4 pm on April 18, 1955. His ashes were scattered at an undisclosed place.

leading to the famous 'twin paradox'. That is the message of the second equation: t' is the time measured by the moving clock, and t the time on the clock standing still. The formula teaches us that t' is always smaller than t (because the square root is smaller than one). If one twin stays home and the other goes on a long journey through space with a high speed, then upon returning home the travelling twin will be much younger. This prediction has been tested in very many instances, though maybe not with a real twin member traveling through space. For example, a very clean experimental proof can be obtained by measuring the decay rate of unstable particles that are produced with a variety of high speeds. At high speed, the particles should live longer, in agreement with Einstein's equation. Similarly, to an observer at rest moving objects appear shorter in the direction in which they move.

Let us now turn to what may be most celebrated formula of twentieth-century physics: $E=mc^2$. In fact, physicists prefer to present it in a somewhat different form nowadays, because m in the Einstein formula is itself a quantity that depends on the state of motion with respect to the observer. What we normally refer to as the mass of an object is the mass we measure if the object is at rest. So 'mass' refers to the rest mass m_0. In the last part of the formula the energy is rewritten in terms of the rest mass and the momentum p (and c), demonstrating two interesting features of relativistic particle dynamics. If we set the momentum equal to zero, the energy is just equal to the rest mass m_0 times c^2, showing that even when the particle is at rest it represents a tremendous amount of energy. To give an idea, it means that in one kilogram of any form of matter about 10^{17} Joules is stored, the same amount of energy one gets from burning more than a million tons of coal.

Maxwell versus Newton

I would like to make an additional remark about the first postulate of relativity. The fact that the laws of nature should look the same for all inertial observers was actually already inconsistent with the two great classical theories: the laws of Newton and the Maxwell equations. There is a rule by which the space and time coordinates of one observer can be expressed in those of the other. This is a so-called 'coordinate transformation', which depends on their relative velocity. If one does this transformation in such a way that Newton's laws read the

Expressing the so-called *relativistic mass m* in terms of the rest mass and the velocity, one obtains an interesting expression, used by Einstein (given in the side bar). This formula has some illuminating features. Observe that if $v=0$, $m=m_0$, as it should be. On the other hand, note that as v approaches the velocity of light, the denominator tends towards zero, and the value of m increases without bound! This does explain to a certain extent why we cannot push the speed of the particle beyond the velocity of light: the particle becomes infinitely heavy, and to accelerate it further would require an infinite force or energy.

$$m = \frac{m_0}{\sqrt{1-v^2/c^2}}$$

same in the new coordinates as in the old, then the Maxwell equations change, and if one chooses the transformation that preserves the form of the Maxwell equations, the Newtonian equations change. It was this deep conflict between the two cherished dogmas of classical physics that was resolved by Einstein's theory; the transformation that preserved the Maxwell equations (the so-called Lorentz transformation) was the correct one, and Newtonian mechanics had to be overthrown, only surviving as a limiting case of the more powerful relativistic mechanics formulated by Einstein.

Another interesting case of the relation between energy and mass (the third equation) arises if the rest mass m_0 is set equal to zero. We then obtain a relation for massless particles, stating that the energy becomes proportional to their momentum: $E=pc$. As we will learn to appreciate somewhat later in the context of quantum theory, it makes a lot of sense to talk about massless particles. In fact, they are all around us in tremendous numbers, in the form of photons or light particles, which are understood as the quanta of electromagnetic radiation. The expression for the energy of massless particles implies that these particles always have to move with the speed of light – you can't put them to rest.

The equations of special relativity make strong statements of a shocking simplicity. They teach us that nature behaves entirely differently if we deal with velocities that are of the order of the velocity of light. The reason we did not discover all this a long time ago is of course that we did not have any experience with such velocities, but nowadays in the great accelerators elementary particles are sped up to a velocity less than one part in a million away from the maximally allowed speed c.

Einstein asked himself what it would mean to have equivalence between accelerated observers. The question led him to a solitary route of about ten years' hard work that took him from Special Relativity to the astonishing theory, prosaically called General Relativity, in which the notions of space, time and gravity got intimately linked.

The phenomenon of gravitation is a direct manifestation of the curvature of space-time. Energy and momentum do curve space-time, and the curvature of space-time affects the motion of matter and radiation. Einstein 'liberated' space-time from its role as the rigid arena in which physics took place. He turned space-time from a passive spectator into an active player, itself taking part in dynamical physical processes. General relativity explained numerous observed facts that Newton's theory could not, it predicted mind-boggling phenomena like black holes and most importantly it opened the door to a completely new, dynamical perspective on our universe.

General relativity
The Einstein equations

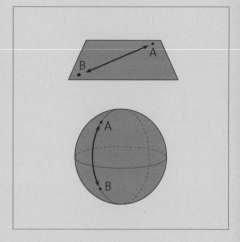

We all have the experience of being in an elevator that is accelerated upward or downward. You know when you are being accelerated upward, because you feel heavier. And if you accelerate downward, you feel lighter. Imagine now that somebody cuts the cables of the elevator so that it would be in free fall. In that elevator everything would be 'weightless': if you would drop something out of your hand, it would not fall down on the floor, but just move along with you. From these observations we may conclude that a theory of relativity for accelerated observers should include aspects of gravity. The freely falling observer would play a unique role because she is weightless; for her there is no gravity. Einstein discovered the key idea in 1907 with the *principle*

$$R_{\mu\nu} - \frac{1}{2} g_{\mu\nu} R + g_{\mu\nu} \Lambda = 8\pi G_N T_{\mu\nu}$$

of *equivalence*, in which gravitational acceleration was postulated to be indistinguishable from acceleration caused by mechanical forces. Gravitational mass, as in Newton's gravitational equation, was therefore identified with inertial mass (appearing in Newton's celebrated formula **F**=*m***a**). Back then this was a phenomenological assertion, but not a logical necessity or a fundamental principle.

There is one other observation to be made in connection with accelerated observers: imagine a light ray. By definition it travels along a straight trajectory between two points, because that is the shortest way to get from one point to another. But if we describe that line in the reference frame of an observer who moves with a constant acceleration, then the light ray would move along a parabola (a curve of constant curvature), and the curvature would be equal to the gravitational acceleration, i.e. to the gravitational field. The shortest path would be curved, and that is the signature of a space that is itself curved. Thus, what Einstein's discoveries amounted to was a relation between acceleration and gravity on the one hand and between acceleration and curved space on the other, so that the identification of gravity and the curvature of space-time was inevitable.

To imagine what a curved space is, it might help to step down from three to two dimensions. Let us compare the surface of a flat plane to the curved surface of a sphere (see the figures on page 64). The latter is actually a surface of constant curvature, because the way the surface is curved is the same at all points and in all directions. On a sphere the shortest path between two points is indeed curved – a segment of a large circle.

Matter causes a gravitational field, from which Einstein concluded that each mass should curve the space-time

Seven predictions

Perihelion precession. The force between two masses, such as the sun and each of the planets, deviates from Newton's 'inverse square' law by some very small terms. Small though they may be, they do have the effect that the long axis of the elliptical planetary motion is no longer fixed in space but itself very, very slowly rotates around the sun. For the closest planet, Mercury, this precession is about 43 seconds of arc per century. The observations are in complete agreement with the theory.

The bending of light. Light rays are affected by the curvature of space-time (i.e. by gravitational force). This is not the case in Newton's theory. Light rays should bend if they pass near heavy objects. This prediction was confirmed during a solar eclipse in 1919. Light from distant stars was bent when passing nearby the sun, causing an apparent displacement of those stars with respect to the other stars. Most spectacular is the effect of around it. In two dimensions a mass would deform the flat space into a plane with a bump around the mass, as if it were a rubber sheet on which the mass is placed.

The Einstein equations express exactly the relations just described. On the left-hand side the $R_{\mu\nu}(x,t)$ fields denote the components of the curvature in the various space-time directions in every space-time point, whereas on the right-hand side we find a multi-component field $T_{\mu\nu}(x,t)$ which describes the density of energy and momentum in space-time. The indices μ and ν are space-time indices; they run over four values 0, 1, 2 and 3, where the zero refers to the time component. So on the left-hand side appears all the information on the geometry of space-time, while on the right-hand side all the information about the distribution of matter and radiation is given. The term with the so-called 'cosmological constant' Λ could have been written on either side of the equation, depending on its interpretation. On the one hand matter and energy determine how space-time is curved, and on the other the curvature of space-time is like the gravitational force and therefore determines how particles but also light move.

The crucial point is not only that space-time is curved, but more importantly that it is dynamical. This means that it is not just a mathematical arena in which physics takes place, but space-time itself is an active player, taking part in dynamical processes just like other physical degrees of freedom.

gravitational lensing, where because of the extraordinary curvature we see multiple images of a far-away galaxy.

Gravitational red shift. Light can lose or gain energy in the gravitational field. This does not mean that it moves more quickly or slowly, but that its frequency will decrease (red shift) or increase (blue shift). This effect played an important role in the discovery of the expansion of the universe.

The expanding universe. The equations that describe the changes in space-time should also be obeyed by space-time as a whole. Einstein's theory determines precisely the dynamics of the universe. In 1922 the Russian mathematician Aleksandr Friedmann published a class of solutions corresponding to expanding universes in the *Zeitschrift für Physik*, and in 1929 Edwin Hubble made the splendid discovery that our universe was expanding indeed.

There is another analogy which might help us grasp this radical turn in our conceptual understanding of gravity implied by the Einstein equations, this time not in the guise

of a curved space-time. In the sections on electromagnetism
we pointed out that the law that specifies the attractive or repelling force between two point charges is very similar to the force law for the gravitational attraction between two point masses as given by Newton. We furthermore saw the emergence of electric and magnetic fields, not just as mathematical constructs to calculate forces between charges, but as independent physical degrees of freedom that had to satisfy their own system of equations, the Maxwell equations. These equations had solutions corresponding to electromagnetic waves propagating through the vacuum – by definition with the speed of light. This resolved the following long-standing puzzle.

Imagine we would jiggle a charge somewhere, then you would expect the field to change and that change in the field could cause another charge very far away to move. According to the old electric force law, the force would be transmitted instantaneously, but if you study the Maxwell equations you would find that the change in the field only propagates with the velocity of light – saving of course the sacred concept of causality.

Now Einstein understood very well that with Newton's law of gravity the same 'instantaneous action at a distance' would cause a conceptual problem. In other words, with special relativity Einstein had resolved the conflict between Maxwell and the dynamical equations of Newton, but the conflict with the gravitational force law had not yet been resolved at that point. He knew what he should look for: a set of equations for a gravitational field that would play the same role the Maxwell equations did for electromagnetism. The equations should be similar but certainly not the same, one reason being that the gravitational force between

Gravitational radiation. The theory predicts the existence of gravitational waves caused by strong gravitational accelerations of massive bodies, in analogy with the electromagnetic waves of the Maxwell theory. This prediction still lacks a direct experimental verification; however, there exists beautiful indirect evidence for the energy loss of a binary pulsar due to gravitational radiation. The measurements for which Joseph Taylor and Russell Hulse received the Nobel Prize in Physics in 1993, which by now extend over more than twenty-five years, are in excellent agreement with the calculations based on the Einstein equations.

Black holes. The theory had yet another surprise in store. It predicted peculiar solutions known as black holes, a prediction that didn't seem to make any sense the time it was first made. In the meantime, there is a growing amount of indirect experimental evidence for their existence. Black holes come into being

69 masses is always attractive. And that is exactly what he found. The fact that his equations at the same time allowed a geometric interpretation implying the curvature of space and time formed an unexpected but magnificent bonus from this point of view.

The Einsteinian theory of gravity belongs to the most esthetical achievements of theoretical physics. It embodied a great conceptual turning point and made a number of remarkable predictions, going very far beyond Newton's theory. The most important ones are listed in the side bar. With this impressive list of predictions, which to a large extent have been vindicated by a variety of experiments, General Relativity stands out as one of the richest physical theories that have ever been conceived.

if a heavy stellar object that has burned up all its nuclear fuel collapses under its own gravitation, also called a supernova. Observations have been made of supernovae where the mass is so large that we can only interpret the event as the formation of a black hole. It is also conjectured that giant black holes sit at the centers of galaxies, slowly gobbling them up.

The cosmological constant. Finally, there is the last term on the left-hand side with the so-called 'cosmological constant' Λ. This term is nowadays interpreted as representing the energy of the vacuum, which actually causes a gravitational repulsion. It has become clear recently that quite unexpectedly, this form of energy dominates the energy density in the universe, causing the universe to go through an accelerated expansion. This underscores the theoretical wisdom that things that are not forbidden by the principles of the theory are 'mandatory' and will eventually show up in nature.

The first quarter of the twentieth century saw two great scientific revolutions: firstly relativity, which deeply changed our view of space and time, and secondly quantum mechanics, which deeply changed the way we think about states of matter and energy. Quantum mechanics provided a deep understanding of a wide variety of phenomena, from the most fundamental properties of elementary particles to many aspects of chemistry. Quantum theory opened the door to the micro-cosmos, thereby explaining many of the properties of vastly different types of materials such as insulators, metals, semiconductors and superconductors. Modern physics is still deeply involved in further exploiting the ever-surprising worlds of quantum solids, liquids and gases.

The quantum revolution implied a radical shake-up of the foundations of theoretical physics. It was caused among others by the inadequacy of the classical theories of Newton and Maxwell to account for the observed structure of atoms. The atom supposedly consisted of a positively charged nucleus surrounded by a number of negatively charged electrons, exactly compensating the nuclear charge. The problem that arose may be explained as follows. Newton told us that in the proposed configuration the electrons would 'circle' very rapidly around the nucleus (like a tiny planetary system), and therefore these electrons would continuously be strongly accelerated. Maxwell, however, told us that an accelerated charge would start radiating and thereby losing energy at an appreciable rate, so that in the end the electrons would fall into the nucleus. A little calculation showed that the lifetime of the atom would be something like a billionth of a second. According to our theoretical knowledge, therefore, atoms would just not exist, bluntly contradicting the observations.

Quantum mechanics
The Schrödinger equation

Erwin Schrödinger

Quantum theory originated from the constant introduced by Max Planck in his radiation law in 1900. Later Niels Bohr used the idea of matter waves for the explanation of atomic spectra. Heisenberg gave an algebraic treatment, called matrix mechanics, while Erwin Schrödinger formulated his wave mechanics. He published the first of four papers containing the celebrated equation in the *Annalen der Physik* in 1927, and showed that the atomic spectra resulted from solving his wave equation for the vibrational states. Schrödinger moved from Zürich to Berlin as Planck's successor. In the year 1933 he shared the Nobel Prize for Physics with Dirac. With Hitler's coming to power, however, Schrödinger was outraged and decided to leave Germany to take up a fellowship in Oxford. In 1934 he was offered a permanent position at Princeton

$$i\hbar \frac{\partial}{\partial t} \Psi(\mathbf{r},t) = \left[-\frac{\hbar^2}{2m} \nabla^2 + V(\mathbf{r}) \right] \Psi(\mathbf{r},t)$$

Quantum theory came to the rescue. Looking back, one of the outstanding achievements of quantum theory is that it explained the structure but also the stability of atoms and molecules, and therefore of all matter. In quantum theory the atomic bound states of the electrons with the nucleus form a discrete set – the allowed states are quantized, with a stable lowest energy state.

The theory is based on a number of postulates which are of great generality, providing a framework that should in principle be applicable to any physical system. To put it briefly: our world *is* quantum-mechanical. The framework of quantum mechanics does not follow from classical physics in any way; it was only through the failures of the classical theories to explain a number of crucial observations that quantum mechanics was conceived step by step. What is true is that under appropriate circumstances the classical laws of physics follow from quantum theory, but not the other way around. The scale at which quantum theory becomes indispensable is very small; it is set by a universal constant called Planck's constant, $h = 6.63 \times 10^{-34}$ Js. The units of h are Joules times seconds, or energy times time. This means that energy can be written as Planck's constant divided by time or – which is the same – multiplied by a frequency.

The central equation of quantum theory, describing the physical states and their evolution in time, is the celebrated Schrödinger equation. In classical mechanics we think of a particle as a point with a given mass, and the state of the particle is specified by its position and velocity. Given the force **F** or a potential $V(x)$ from which the force can be calculated, Newton's equations tell us how the particle will

University, but he did not accept it. He took a position at the Institute for Advanced Studies in Dublin in 1938, where he stayed till his retirement in 1955.

Schrödinger was not happy with the statistical interpretation of his wave function, and tried to set up a theory in terms of waves only. This led him into a controversy with other leading physicists. After his retirement he returned to Vienna. In 1944 he published the little book *What is life?* in which he presented a molecular and quantum-mechanical view on the problem of life. It became a source of inspiration in the emerging field of molecular biology. Schrödinger worked alone, only two or three of his ninety papers were written together with others. He died in January, 1961, after a long illness.

move, i.e. how its position and velocity will change in time.

In quantum theory the situation is entirely different: the state of the system, say a particle, is described by a wave function $\Psi(x,t)$ and the allowed states and their dynamics are governed by a differential equation that this wave function has to satisfy.

For a free particle on which no force is exerted, the Schrödinger equation reduces to a free wave equation. This expresses the fact that in quantum mechanics particles are treated as waves and vice versa. In the early days, quantum mechanics was indeed often called 'wave mechanics'. The basic feature referred to as *particle–wave duality* is in its most elementary form expressed by two simple relations. One was due to De Broglie, relating the wavelength λ of the 'matter wave' to the momentum of the particle (as $\lambda = h/mv$), and the other was due to Einstein, relating its frequency to the energy of radiation ($E = hf$). Note that Planck's constant plays a crucial role in these relations.

The smallness of h in the particle–wave duality relations makes clear why the quantum-mechanical nature of matter remained hidden for such a long time. If we take $m = 1$ kilogram and $v = 1$ km/second, the corresponding wavelength would become $\lambda = 10^{-34}$ meter, which is of course exceedingly small and therefore undetectable. But if one considers an electron with a tiny mass of $m = 10^{-30}$ kg, which moves with appreciable velocity, then the wavelength may become of the order of 10^{-10} meter, which is typically the size of an atom. In other words, if we want to study the behavior of electrons in an atom, we have to use a quantum-mechanical description, because the wavelength of the electron 'wave' is of the order of the size of the system, and typical wave properties such as interference will be important.

Photons: the particles of light

If particles can manifest themselves as waves, is it then also true that waves can act as particles? This is what the Einstein relation $E = hf$ expresses. If we consider light of a certain color (= frequency), then this could be considered as a stream of massless particles (or quanta) called *photons*. And indeed, De Broglie's and Einstein's relations agree if we use the relation $E = pc$ (and $c = \lambda f$) for massless particles, as mentioned in the section on special relativity. The conclusion is that quantum theory to a certain degree unifies the completely different classical concepts of particles and waves.

Indeed the first great success of the Schrödinger equation was that the spectrum of atomic states could be calculated with great precision. It showed that the system of an electron and a nucleus had a stable lowest energy state; a theory had been obtained explaining and underlying the stable elements in the periodic table of Mendeleev.

Given the wave function, what do we know and how do we recover the properties of the physical system? The most common interpretation of quantum mechanics is to say that the magnitude of the wave function squared is nothing but the *probability* to find the particle at location **r** at time t. This appears somewhat paradoxical. Imagine a solution which is an honest sinusoidal wave. It has a well-defined wavelength, and because of the De Broglie relation also a well-defined momentum. However, at the same time the particle is not very well-localized, because the wave function is virtually non-zero everywhere. The other extreme corresponds to a situation where the wave function is very well-localized and highly peaked around a single point. This wave function does typically describe a particle localized at that point, but now this function does not have a well-defined wavelength; i.e. we cannot say much about its momentum or velocity.

Now this implies a surprising statement that quantum theory makes about physical reality. One cannot at the same time know everything about the position and velocity of a particle. Particle–wave duality implies an uncertainty in position $\triangle x$ and an uncertainty in momentum $\triangle p$, which together have to satisfy the Heisenberg uncertainty relation given in the side bar: their product has to be *at least* the small number $h/4\pi$.

In this sense, quantum mechanics seems a retrenchment

$$\triangle x \triangle p \geq h/4\pi$$

from the ambition and expectation of classical physics, where nature was completely predictable at the basic level. Predictability would require the knowledge of both the position and the momentum of a particle to an arbitrary accuracy, and that is exactly what the uncertainty relation excludes as a matter of principle. If one knows exactly where a particle is at some time but one doesn't know anything about its velocity, clearly one cannot tell where that particle will be a short time later. Because of the Heisenberg relation, one should speak of a 'certain uncertainty'.

This probabilistic aspect of quantum mechanics has given rise to fierce debates ever since its conception. In particular, the uncertainty implied by the theory has sometimes been perceived as marking a fundamental incompleteness of the theory. It was suggested there might be *hidden variables*, but these have been excluded by a theorem by John Bell and the experimental verification thereof. On a more philosophical level, the overwhelming experimental evidence boils down to the statement that we have to accept the probabilistic aspect as fundamental in nature, implying that certain questions – like what the velocity of a perfectly localized particle is – just do not make sense. Up to now quantum theory, with all its counter-intuitive aspects, is probably the most successful theory of nature we know. As we shall see in the remaining part of this book, its postulates hold all the way down to the deepest known levels of nature.

Position–momentum duality

The uncertainty relation can be understood in an analogy with sound waves. A sound wave with a single frequency (or wavelength) corresponds to a pure tone, which is similar to a particle state with a precisely fixed momentum. To accurately determine the height of the tone, we have to hear it over a certain amount of time, which extends over many oscillations. This implies that a pure tone is not very well 'localized in time'. If in contrast I clap my hands, the sound is very brief and thus very well-localized in time, but if I want to know what tone it corresponded to, then the correct statement is to say that all frequencies or tones would be present in the sound. Apparently, you can't have both.

A serious limitation of the Schrödinger equation is that it is not compatible with relativity. The Dirac equation solved that problem; it unites the concepts of quantum mechanics and special relativity, describing the quantum properties of particles like electrons, protons, neutrinos and quarks. The analysis of this equation elegantly explained some of the more elusive particle properties like spin, and provided a solid foundation for the so-called Pauli Exclusion Principle needed to explain atomic structure and the periodic table. Last but not least, the equation predicted the existence of antimatter: the fact that for any particle 'species' there exists an associated species with exactly the opposite properties (such as charge), but the same mass.

The relativistic electron
The Dirac equation

In spite of its tremendous successes, the Schrödinger equation had a serious drawback: it was not compatible with special relativity. This may be inferred from the fact that in the equation the space and time variables x and t do not appear on equal footing: it contains a first derivative with respect to time, but a second derivative with respect to the spatial coordinates. Dirac solved this problem with the equation carrying his name.

The Dirac equation has quite an involved mathematical structure, which is somewhat hidden by the compact notation, so let us take some time to comment on the notation used. There is an index μ which can take the values 0, 1, 2 or 3, indicating time and the three space components, indeed appearing on equal footing. The four A_μ fields, called 'electromagnetic potentials', describe the electromagnetic field in which (for example) the electron moves, and m_e is the electron mass. The electron field is here described by a four-component function Ψ. The so-called 'gamma matrices'

Paul Adrien Maurice Dirac

Paul Adrien Maurice Dirac was born on the 8th of August 1902 in England, his father being Swiss and his mother English. He studied electrical engineering and mathematics in Bristol and attended St. John's College, Cambridge, as a research student in mathematics. He received his doctoral degree in 1926 and became a Fellow of St. John's College. He was elected a Fellow of the Royal Society in 1930, and became Lucasian Professor of Mathematics at Cambridge in 1932. Dirac shared the Nobel Prize in Physics with Schrödinger in 1933. In 1971 Dirac took up a professorship at Florida State University in Tallahassee. He died in 1984.

$$\left\{ \gamma^\mu \left(i \, \frac{\partial}{\partial x^\mu} - eA_\mu \right) - m_e \right\} \Psi(x^\nu) = 0$$

γ^μ are four numerical matrices (4x4 arrays of given numbers) which have to be multiplied in a standard mathematical way with the components of Ψ. (Actually, we have suppressed an extra component index on Ψ to prevent the notation from becoming even more involved).

Analysis of the equation revealed the meaning of the four components of the Dirac field. It includes the description of the somewhat mysterious property called *spin*, best described as some intrinsic rotational degree of freedom. We could say that the electron is the quantum equivalent of a tiny spinning top – and it can be left- or right-handed.

Remarkably, the equation turned out not just to describe the two spin components of an electron, but also the two spin states of another particle with exactly the same mass but the opposite (positive) charge. This particle is therefore called the *positron*. C.D. Anderson discovered this first example of an 'antiparticle' experimentally in 1932. It became clear that in fact all particles in nature have antiparticles, with exactly opposite properties such that when a particle and an antiparticle meet, the pair can annihilate each other and be converted into pure energy in the form of electromagnetic radiation – a dramatic instance of the equation $E=mc^2$. Because of the peculiar spin properties, the four-component object Ψ is called a *spinor* rather than a vector.

A further analysis of the Dirac equation also led to an explanation of Pauli's *exclusion principle*. This rule, obeyed by electrons and all other particles described by a Dirac-type equation, decreed that two or more of these particles could never sit in exactly the same state. This was a crucial but up to then ad hoc ingredient of quantum theory, needed to explain the periodic table of atoms. Indeed, as the electrons in an atom cannot all sit in the same lowest energy state, they

Folklore has it that Dirac hated publicity and was a man of few words. However, his book on quantum mechanics is still one of the most elegant and well-written on the subject. Maybe both fit his statement: 'I was taught at school never to start a sentence without knowing the end of it.'

have to systematically fill up the higher energy levels, causing different types of atoms to display entirely different chemical behavior.

Quantum Electrodynamics

The Dirac equation together with the Maxwell equations constitutes Quantum Electrodynamics (QED), the quantum theory of electrons, positrons and photons. This theory was completed after the Second World War by the American physicists Richard Feynman and Julian Schwinger and the Japanese Shin-Ichiro Tomonaga. It has been tested to extreme accuracy, for example by a variety of very precise measurements of the magnetic moments of the electron and a particle called the *muon*. Quantum electrodynamics is the prototype of what is now called a *quantum field theory*, the framework in which the relativistic dynamics of elementary particles and fundamental forces is nowadays formulated. It provides an extremely successful description of how nature behaves on the most basic level.

It is remarkable that in this framework the distinction between forces and the particles on which these forces act is removed; quantum fields describe both. A force is described as the exchange of some intermediate particle that 'carries' that force; for electromagnetism this is the well-known photon. Only after the development of quantum field theory could one hope for a comprehensive treatment of all fundamental interactions in nature from a unified perspective.

Quantum fields

In classical physics we deal with particles, forces and waves. In quantum theory we discovered that somehow the very different classical notions of particles and waves became unified in the sense that there appeared to be complementary. In quantum field theory we add special relativity to the quantum picture, which means that the equivalence of mass and energy should be implemented in the theory. It does: the states of a quantum field theory are multi-particle states that describe different numbers of different particle types, each of them satisfying the Einstein relation between energy mass en momentum. The quantum fields can be used to create or annihilate particles so that interactions can be taken into account. It is a language in which all particles and forces are described on equal footing as quanta of the corresponding quantum fields.

To our present knowledge, there are four fundamental forces in Nature. We have already encountered gravity and electromagnetism, because they manifest themselves directly in the macroscopic world, and we have displayed the fundamental equations describing them. However, other known facts could not be explained by these two forces. For example, it was discovered that the nuclei of atoms were composed of smaller building blocks denoted *neutrons* and *protons*, the latter being positively charged. This raised the following question: if equal charges repel, how come these positively charged protons can all sit so peacefully together within the nucleus? Why doesn't the nucleus fly apart? The answer was simple: there is a force stronger than the electromagnetic force – properly referred to as the *strong force* – which keeps the nucleus together and which works on protons and neutrons in the same way. The theory describing this force is called Quantum Chromodynamics (QCD). During the 1970's it was discovered that the protons and neutrons themselves were composite particles too, each containing three so-called *quarks*. These quarks have never been observed as free isolated particles, because apparently they are permanently bound to each other and confined to the inside of particles called *hadrons*, like the proton and neutron. The quarks are subject to the strong force because they carry a kind of charge called 'color' (which has nothing to do with ordinary color), which keeps them tightly bound in these composites.

Quantum Chromodynamics describes the behavior of quarks and the strong force. This force is mediated by particles called *gluons*, because they 'glue' the quarks together in colorless composites called hadrons. There is a residual strong force

The strong force
Quantum chromodynamics

Historical note

Murray Gell-Mann and George Zweig first proposed the idea of quarks in the 1960s, but it was not until ten years later that quantum chromodynamics was formulated as a theory for the strong interactions between quarks, binding them into protons and neutrons. The early proposals were due to Yoichiro Nambu and to Murray Gell-Mann and Harald Fritzsch using equations already postulated in a different context by Yang and Mills in 1954. Crucial to the successful predictions of the theory was the discovery of asymptotic freedom – the property that the strong interaction becomes weak at short distances. For this discovery the 2004 Nobel Prize in Physics was awarded to David Gross, David Politzer and Frank Wilczek. The property of asymptotic freedom lies at the basis of many attempts to further unify the description of the fundamental interactions.

$$\mathcal{L}_{QCD} = -\frac{1}{4} F_a^{\mu\nu} F_{\mu\nu}^a + \sum_f \bar{\Psi}_f (i\partial\!\!\!/ - M + g_s A\!\!\!/^a T_a) \Psi_f$$

$$F_{\mu\nu}^a = \partial_\mu A_\nu^a - \partial_\nu A_\mu^a + g_s f_{bc}^a A_\mu^b A_\nu^c$$

between the hadrons, which for example binds neutrons and protons together in the nucleus. This is very similar to atomic physics, where the atoms are held together by the electromagnetic force, which is also responsible for the binding of atoms into molecules or other structures like crystals. Unlike quarks, electrons do not carry color and are therefore insensitive to the strong force. Hence they are not confined.

Quantum chromodynamics is completely specified by the first formula given. The second equation gives the definition of colorfields F in terms of the potentials A. The formulas are of remarkable beauty but deceiving simplicity. This is due to a particular type of mathematical symmetry called 'gauge symmetry', which is hidden in the formulation of the theory.

We briefly summarize some of the basic features of the expression, without getting too specific. The first term, quadratic in F, describes the gluons (labeled by the index a) and the strong interactions among them. Whereas the electromagnetic force is mediated by one type of particle only, the photon, in quantum chromodynamics we have to deal with eight different particles, the gluons. The gluons satisfy a set of equations called the Yang–Mills equations, first written down by Chen Ning Yang and Robert Mills in a different context already in 1954. These equations form a beautiful generalization of the Maxwell equations, exploiting the symmetry principle denoted as local gauge invariance. Never mind the elegance of the equations, because what makes life particularly difficult is that the gluons themselves carry color charges and therefore exert strong forces on each other, so that they also are confined.

The second term, with the two Ψ fields, describes the six different quark types (labeled by index f because they are often

The Lagrangian

How is it that we can characterize a complete theory by just a single formula? **L** is called the *Lagrangian*, after its inventor Joseph-Louis Lagrange. It is a function of fields that is closely related to the expression for energy. Whereas we usually write the total energy E as the sum of a kinetic part (related to the motion) and a potential part, the function **L** is the difference of the kinetic and the potential part. The reasons for specifying the theory by this function are its built-in symmetries and its economy – a relevant aspect as the theories we consider become more and more involved. From the function **L** one can directly derive all the coupled equations that the different fields have to satisfy, including their interactions. We could also have given the Lagrangian for quantum electrodynamics, which amounts to specifying both the Maxwell and Dirac equations for electromagnetic fields.

called quark flavors). It includes a part with the A^a fields, determining the interactions of the quarks with the gluons. The quark equations are like the Dirac equation with the electron fields replaced by the quark fields and the photon field replaced by the gluon fields. The superficial similarity between the two theories is not an accident: it reflects a universal feature of how the forces in nature work. Yet the complicated dynamics of the gluons among themselves make the combined dynamics of quarks and gluons – and thus of nuclear particles – difficult to extract from this theory.

There is a miraculous property, however, which in the end gives us a handle on some crucial aspects of the strong interactions. It is denoted as *asymptotic freedom*. It turns out that if we probe very deeply within the proton and study what happens to the strong force when the quarks get very close to each other, we learn that the color force becomes ever weaker. The reason for this is basically that since the gluons themselves carry color charge, the charge on the quark effectively spreads out over a cloud of gluons surrounding the quark.

Now, how can we characterize the strength of this force? Is it strong or weak, or both at the same time? The strength of the gravitational force is specified by Newton's constant, for electromagnetic force the fundamental electric charge comes in, but what can we say about the coupling parameter g_s for the strong force? The thing to do is to specify its strength at a given scale – a scale of length, energy or mass (because these are all linked through the universal constants h and c). One could say that the scale where the interaction strength becomes considerable (say of order unity) sets the scale of the physics of this theory. That scale is the mass of the proton, m_p. And this is why the proton mass is given in the list of fundamental particles.

Asymptotic freedom

The property of asymptotic freedom can be understood from an analogy. It is similar to the way the strength with which the earth attracts a massive object would vary if that object would move *through* the earth towards the center. One can show that when the object is sitting at a certain radius, only the mass of the part of the earth inside the sphere with that radius contributes to the central force. Therefore, going ever deeper towards the center, the 'effective mass' of the earth would decrease, and the strength of the gravitational force would tend to zero at the center. In a similar way the strong force becomes weak at small scales, and the quarks behave like free particles. Thanks to asymptotic freedom, predictions could be made that have been splendidly confirmed by experiments in many of the great accelerator centers around the world.

The last of the four known forces is the *weak force*. It is responsible for the spontaneous disintegration of certain radioactive nuclei – like uranium – where a neutron in the nucleus decays into a proton, an electron and an antineutrino. The weak force is thus the force behind fission energy. Becquerel and the Curie couple discovered such decays already in 1896, but it took until about 1970 before a satisfactory theory, compatible with the principles of quantum mechanics, was completed.

The weak force acts upon quarks as well as electrons, neutrinos and similar particles. Most remarkable is the fact that the consistent theory also includes the electromagnetic interactions. A forced but nevertheless exuberant marriage between both interactions was achieved. The electro-weak theory is described in terms of four particles (three W and a single B particle) that mediate the forces, three of which have to do with the weak force (usually denoted as W^+, W^- and Z particles), while another combination of them is to be identified with electromagnetism (the photon, denoted as γ).

The defining expression looks horribly complicated, and yet it is not so hard to recognize the same generic structure as in other examples discussed before. The total Lagrangian L_{E-W} is composed of four parts. The first part L_g describes the particles carrying the forces and the interaction between them (completely analogous to the F squared term for the gluons in the equation for the strong force). The second part L_f contains all the Dirac-type particles and their interactions with the force-carrying particles. The label i indicates the various particle species – electrons, quark flavors, neutrinos, etc. There is an important subtlety, which has to do with the fact that the weak interactions work differently on the left-

Electro-weak interactions
The Glashow–Weinberg–Salam model

Historical note

The electro-weak theory was formulated in 1968, but its predictions could only be verified by accelerator experiments much later. In 1993 Shelly Glashow, Steven Weinberg and Abdus Salam shared the Nobel Prize for proposing the theory. The W particle was discovered at CERN, for which Carlo Rubbia and Simon van der Meer received the Nobel Prize in 1984. The proof of mathematical consistency of the theory as well as the development of crucial calculational tools was done by Gerard 't Hooft and Martinus Veltman. This made it possible to make very precise predictions, which have been confirmed by numerous experiments at the large accelerator centers around the world. They received the Nobel Prize for their work in 1979.

$$\mathcal{L}_{E-W} = \mathcal{L}_g + \mathcal{L}_f + \mathcal{L}_H + \mathcal{L}_m$$

$$\mathcal{L}_g = -\frac{1}{4} G_a^{\mu\nu} G_{\mu\nu}^a - \frac{1}{4} B^{\mu\nu} B_{\mu\nu}$$

$$\mathcal{L}_f = \sum_i \bar{\Psi}_{Li} (i\partial\!\!\!/ + g' W\!\!\!\!/\,^a t_a + g B\!\!\!/ y) \, \Psi_{Li}$$

$$\qquad + \sum \bar{\Psi}_{Ri} (i\partial\!\!\!/ + g B\!\!\!/ y)_{Ri}$$

$$\mathcal{L}_H = - (D_\nu \phi)^\dagger (D^\nu \phi) - \mu^2 (\phi^\dagger \phi) + \lambda (\phi^\dagger \phi)^2$$

$$\mathcal{L}_m = - \sum_{i,j} \left(c_{ij} \bar{\Psi}_{Li} \phi \Psi'_{Rj} \right)$$

handed (*L*) and on the right-handed (*R*) spin component of
the Dirac-type particles. In other words: nature breaks mirror
symmetry on a very fundamental level, preferring the left-
handed variety of matter. This means that if a certain process
may take place, the mirror image of that process may turn out
to be impossible.

In the table of the Standard Model in the side bar, we see
quarks (in three colors) and leptons, six types of each. The six
quark types *up*, *down*, *charm*, *strange*, *top* and *bottom* are the
'flavors' referred to before. The six leptons are in two columns:
the left one contains the *electron* and its heavier copies, the
right column gives the associated *neutrinos*.

So altogether, in spite of the remarkable reduction in the
number of fundamental building blocks of matter, there is
still quite some name-dropping going on. Yet ordinary matter
is basically all built up of particles in the first row (up and
down quarks, and electrons). The three rows are denoted
as the three families of elementary particles, sharing pretty
much the same properties – only their masses increase. The
weak interactions allow the heavy particle species to decay
into the lighter particles of the first family in the top row: the
particles from which the stable matter in our universe is built.
These transitions between the fundamental particles caused
by the weak force are what we observe as the radioactive
decay of nuclei. The strength of the weak force, determined
from these decays, is basically set by the mass scale of the W
particles.

The third part of the Lagrangian is the scalar part L_H. It
describes yet another particle that is instrumental in this
theory, the so-called Higgs particle, which is needed to give
mass to most of the other particles. The Higgs particle has not

THE STANDARD MODEL

Force carriers				Higgs
g	γ	$w^{\pm}$	z	φ

Quarks		Leptons	
u	d	e	v_e
c	s	μ	v_μ
t	b	τ	v_τ

yet been observed, but its discovery is expected to take place when the Large Hadron Collider, a new accelerator at the CERN in Geneva, is completed. The Higgs field is very relevant because in some sense it explains the origin of mass. This field is present as a constant background through which the other particles have to move, thereby acquiring what we call a mass. The mass of the Dirac particles is generated through the interactions given by the last part, L_m.

In spite of all the complications and subtleties, it is gratifying to see the obvious similarities in the theories describing the weak, electromagnetic and strong forces, which are indeed based one and the same gauge symmetry principle. The combined theory for these three forces is called the Standard Model of Elementary Particles. In the table on page 86 all the ingredients are indicated. This is the modern version of the periodic system of elementary particles and entails a dramatic leap forward from the good old periodic table of Mendeleev, which we all once stared at in our high school classrooms. It is paradoxical that exactly the road of uncompromising reductionism has led us to such a highly unified perspective on the whole of nature.

On the fundamental level we have now identified four forces: the Standard Model describes three of them and the fourth – gravity – is described by General Relativity. The Standard Model is a quantum theory, while General Relativity is not. It is not surprising that with these successful steps towards unification in the description of particles and forces, the hope and expectation are that we are heading towards an all-encompassing theory of nature, including a quantum theory of gravity. This is considered the Holy Grail of fundamental physics.

At the moment the most ambitious attempt to formulate such a theory is string theory, in which the fundamental constituents of matter and all forces between them are manifestations of an underlying dynamics of strings. These strings are so tiny that there is little hope they will ever be observed directly. Credibility of the theory has to be achieved by indirect means. Predictions (or outcomes) of the theory are, that all elementary particles have supersymmetric partners, and that the world on a fundamental level would have ten, possibly eleven dimensions. The scale at which these features manifest themselves is still open to debate.

The quest for a unified field theory dates back to Einstein, who spent the later years of his life on this problem. Over the years many attempts have been made, with a varying degree of success. Examples are the highly symmetric Grand Unified Theories, which try to unify strong, weak and electromagnetic forces in a single model. These theories explain for example the simple experimental fact – so far never accounted for by theory – that the electron and proton have exactly the same but opposite electric charge. Another remarkable generic prediction is that the proton will decay (though extremely

String theory
The superstring action

Historical note

String theory first appeared in the early 1960s as a possible theory for the strong interactions, but was abandoned when quantum chromodynamics came along. The theory found its true vocation in 1983 as a serious candidate for a theory of quantum gravity, in fact as a theory of all interactions and all particles. Great contributions to the subject came from many people: we mention the early contributions of Joel Sherk, John Schwarz, and Michael Green and Alexander Polyakov. Edward Witten, presently occupying Einstein's chair at the Institute for Advanced Study at Princeton, made many lasting contributions to the mathematical formulation as well as to the interpretation of the theory. Not surprisingly, string theory in its all-embracing ambition attracts large numbers of talented young researchers from all over the world.

$$\mathcal{L}_S = -\frac{1}{4\pi\alpha'}\sqrt{|g|}\left(g^{\alpha\beta}\partial_\alpha X^\mu \partial_\beta X_\mu - i\bar{\Psi}\slashed{\partial}\Psi\right)$$

rarely) into lighter particles, implying that ultimately all matter would be unstable. Fortunately, the lifetime of the proton is more than 10^{32} years. Nevertheless, these theories would no doubt gain great respectability once such a decay would be observed.

Some people have started from the gravity side. It turned out that the methods of quantum field theory, which yielded a successful quantum description for the other interactions, failed miserably when applied to general relativity. The uneasy feeling grew that maybe Einstein's theory had to be replaced, or at least be generalized to make it consistent with the quantum postulates. The need for such a theory becomes evident if we want to study physics under extreme conditions. The scale for these conditions is set by the universal constants which have a bearing on the different realms of physics we try to bring together: gravity (G_N), relativity (c) and quantum theory (h). Such extreme conditions occur in situations like the early stages of the big bang or the final stages of black hole evaporation.

A first daring and promising revision of Einstein's theory attempting to bridge the gap was the idea of supergravity, where space is extended to some superspace with an underlying supersymmetry. Supersymmetry requires that ultimately all particle types would have their own superpartners. So far, none of these partners has ever shown up in experiments, but that does not mean that one day these divorcees may not show up in court.

The most far-reaching and perhaps robust candidate for a 'theory of everything' is the theory of superstrings. The basic string postulate is that the most fundamental entities in

The ultimate structure of space and time
What does it mean that space-time is made up of strings? Space-time is a collective manifestation of strings, a classical background of the massless 'graviton' modes of the string. As such space and time is an emergent phenomenon in string theory: beyond a certain scale the notion of space-time looses its meaning and questions about what exactly happens should be answered by looking at the strings themselves. Just like if you want understand water on ever finer scales you would end up studying the individual molecules which do no longer have the properties of the collective we call water. So beyond a certain scale it presumable doesn't make sense anymore to talk of space or time.

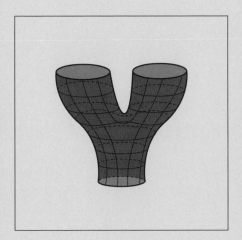

nature are not point-like particles but tiny one-dimensional strings, with a typical size of 10^{-35} m. String theory asserts that each elementary particle type corresponds to a different oscillatory mode of the superstring. To put the extremely small size of the strings in perspective, we note that with the largest accelerators we can now get down to maybe 10^{-20} m, so there is no hope to verify the string conjecture directly. Scientists try to formulate the theory accurately and then see whether falsifiable predictions can be made that indirectly provide evidence for the theory. The starting point is rather straightforward: one postulates that the fundamental entities of nature are strings, formulates an expression for the energy or the appropriate function **L** for them, and then applies the quantum principles to see what physics one gets. One immediately finds some rather exciting results. The lowest states correspond to massless particles, among them the *graviton*, the quantum carrying the gravitational force. It is even possible to derive Einstein's theory as a low-energy, long-distance approximate description of string theory. Also, the particle types and typical interactions in the Standard Model do appear. But another unavoidable and not quite anticipated feature of the theory showed up: the dimensionality of space-time is not our cherished and observed four, but ten! To fit our world into this theory, therefore, six of the ten dimensions have to disappear, or at least become invisible. Six dimensions have to be 'rolled up' like a carpet and be very, very small, so that we cannot observe them. The theory does not tell us how this is to be done. There is evidence that there are very many ways in which it can be done, all of them with certain advantages and disadvantages.

The hope and expectation are that such a theory would make it possible to quantitatively relate observed quantities to each other – for example, that one could calculate the masses of particles or the strengths of forces – but this is extremely hard to achieve because of the huge difference in scales between observationally accessible parameters and the string scale.

The formula we have given defines superstring theory. The first term basically gives the area of the surface (or world sheet) which the string sweeps out while moving through space-time. This surface can be closed or open, in which case the boundaries are closed one-dimensional loops that can be thought of as incoming or outgoing strings. In this geometric formulation the string interactions are already included (see figure on page 91). Looked at in another way, string theory is in some sense a two-dimensional field theory defined on the 'world sheet', in which the space-time coordinates are the fields. This does indeed lead to a quantization of space-time in a very literal and unexpected way.

The second term in the string-Lagrangian contains Dirac-type fields, also propagating on the world sheet. These are necessary to implement the supersymmetry, which in turn is all-important for the consistency of the theory as a whole.

We conclude by saying that the equations of string theory do not have the same status as most of the other equations presented in this book, but many consider this theory as the emerging all-embracing paradigm of nature.

Branes and M-theory

Strings come in two basic varieties: open and closed strings. Closed strings describe gravity, whereas open strings are often linked to the other interactions. The modern version of string theory, often called M-theory, involves not only strings but also higher dimensional objects called p-branes. A 1-brane is a string, a 2-brane is a membrane, higher p-branes are thus higher dimensional analogs of membranes. The rule is now that open strings have to end on branes, in other words open strings can connect p-branes. This is a major amendment to the original string proposal, but is necessary to make the theory mathematically consistent. The whole collection of things is called M-theory where the M stands for Mother, Mystery or more down to earth for Matrix.

Back to the future

A final perspective

$$l_{Pl} = \sqrt{\frac{\hbar G_N}{c^3}} \cong 10^{-35} \ m.$$

$$t_{Pl} = \sqrt{\frac{\hbar G_N}{c^5}} \cong 10^{-43} \ sec.$$

$$m_{Pl} = \sqrt{\frac{\hbar c}{G_N}} \cong 10^{-8} \ kg.$$

Now that we have completed our rather abstract journey at high altitude, let us take another look at the table with the universal constants of nature, given on the inside of the front cover and mentioned at the beginning of the book. It is just a bunch of numbers. We advocated the system of units based on the meter, the kilogram, the second and the Coulomb as units for length, mass, time and charge, respectively. These units are very much tied to the typical human scale, which – as we have been discussing in this book – is rather arbitrary in the grand total of things.

One may wonder whether nature itself doesn't have a favored set of units. Does it? In fact, around 1900 Max Planck already pointed out that a particular combination of the fundamental constants of nature could be used to define a 'natural' system of units of space, time and mass instead of the standard one. The Planck units involve the fundamental constants of gravity (G_N), of relativity (c) and of quantum theory (h). These are called the Planck length, time and mass, and their definitions are given in the side bar. These are the fundamental scales that are intrinsically encoded in the data we retrieved from our physical world.

There is another evocative way to characterize today's grand picture of physics, using an imaginary magical cube positioned in the 'space of theories' (see figure on page 94). Looking at this magical cube, one sees the network of equations we have been discussing in yet another perspective. It is a different – three-dimensional – representation of the 'map of contents' presented at the beginning of the book. The great turning points in physics are positioned at the corners of this cube. At the lower back edge, labeled by Newton's constant, we find the Newtonian physics (mechanical and gravitational theory)

we have been harassed with in high school. The Theory of
Special Relativity connected to the universality of the velocity
of light opened up a new dimension, as symbolized by the
bottom plane coming out of the page. But as explained before,
the Special Theory was still in conflict with the Newtonian
theory of gravity, a conflict that was eventually resolved by
the General Theory of Relativity, where gravity was identified
with the curvature of space-time.

Subsequently, we went through the quantum revolution
linked to Planck's constant, which adds yet another dimension
to our conceptual framework, perpendicular to the space-
time-gravity plane. Back in the upper left-hand corner lives
the Schrödinger equation. The upper-right hand corner in
the back would refer to a non-relativistic treatment of the
gravitational force. This point lies not in physically interesting
domain of parameter space. For example it would represent
the quantum mechanical treatment of the moon-earth or
the planetary system. The parameters would be such that the
result would not differ by any ever-observable effect from the
classical Newtonian result. We then had to face the fact that
now ordinary quantum theory and special relativity didn't like
each other. This conflict was resolved by the development of
quantum field theories like quantum chromodynamics and
the electro-weak theory. So in the front left top corner we
basically see the Standard Model. To put it differently, our
theories of matter live in the left vertical plane. And once
more we encounter a conflict: it is the conflict between our
quantum theories of matter on the one hand and Einstein's
geometric theory of space-time and gravity on the other. This
conflict is basically in the domain of nature parameterized
by the Planck scales, and therefore impossible to probe
directly by experiment. One may hope that string theory will

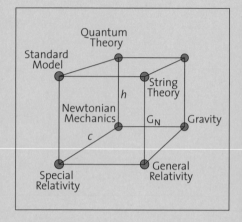

ultimately be able to resolve this conflict hiding so deeply under the surface of everyday experience; string theory as a quantum theory of space-time.

What about the other universal constants, what do they express or encode? We have encountered four fundamental forces, and now it is clear that the strength of each of them somehow defines a typical scale at which nature has organized itself. For gravity, it is Newton's constant. Translated into an energy scale it yields the Planck mass, which as we have seen sets the scale for the fundamental strings. The strength of the electromagnetic coupling, e, translates into the energy scale of atomic physics, determining the typical energy differences between atomic states. These are proportional to the rest energy of the electron, $E_e = m_e c^2$ times a certain dimensionless quantity depending on e, (the square of the *fine structure constant* $\alpha = e^2/2hc = 1/137.0$). The scale for the mass of the atom as a whole is set by the mass of the nucleus and that in turn is set by the mass of a proton, m_p. The proton mass corresponds to the intrinsic scale of the strong force, as pointed out before. It is the energy scale at which the strong interactions become of order unity. So now we are left with the weak force, and its scale is linked to the time-scales of the decay processes caused by the weak interaction. These processes can be directly linked to the mass of the intermediate W-particles.

Looking again at the cube, these constants are scattered somewhere in the left vertical plane, spanned by the h and c edges. This somewhat casual remark may well have some deeper, unexpected content, namely that given the Planck units as the most fundamental ones, the other parameters might not be truly fundamental. Perhaps they are just

Units

The constants of nature have certain values, which depend on the units one chooses to work with. The value of the velocity of light is a different number, of course, depending on whether one talks about miles per hour, meters per second, or feet per moon cycle, for that matter. It is however very convenient if everybody uses the same system of units – it doesn't really matter what system, as long as everybody uses it. The international scientific community has agreed on using the meter-kilogram-second system, including the Coulomb as fundamental unit of charge and the Kelvin as the unit of temperature. The definition of these constants is related with certain quantities that can be measured with great precision, so that these definitions do indeed change in the course of time, depending on the state of experimental art.

dynamical consequences of an underlying fundamental
theory, like the other constants we have encountered – such
as Boltzmann's constant defining the temperature scale in
terms of energy, or Avogadro's constant, which originates
from Amadeo Avogadro's correct hypothesis, made around
1810, that the number of particles in any kind of gas with a
fixed volume and the same pressure and temperature should
be equal. The universal constants listed on the inside of the
front cover set the basic scales that are characteristic for our
world. In specific media, other relative scales may enter the
discussion and generate other phenomenologically relevant
parameters. These are effective parameters dynamically
generated by a subtle interplay of nature's universal laws
and constants. If our observations ever confirm superstring
theory, we expect that only a single fundamental constant
should survive.